Venko N. Beschkov, Dragomir Yankov (Eds.)

Downstream Processing in Biotechnology

Also of interest

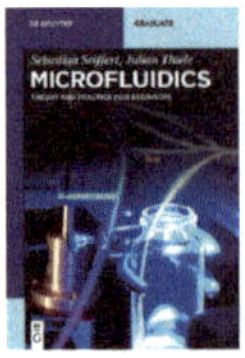

Microfluidics.
Theory and Practice for Beginners
Seiffert, Thiele, 2020
ISBN 978-3-11-048777-0, e-ISBN 978-3-11-048770-1

Cellulose Nanocrystals.
Emerging Bio-Precursors for Chemical Processes
Katiyar, Dhar, 2020
ISBN 978-3-11-064452-4, e-ISBN 978-3-11-064801-0

Pulp Production and Processing.
High-Tech Applications
Popa (Ed.), 2020
ISBN 978-3-11-065883-5, e-ISBN 978-3-11-065884-2

Keratin-based Materials
Reddy, Zhou, Ma, 2020
ISBN 978-1-5015-1913-0, e-ISBN 978-1-5015-1176-9

Environmental Chemistry.
Principles and Practices
Rihana-Abdallah, Benvenuto, Roberts-Kirchhoff, Lanigan, Evans, 2021
ISBN 978-3-11-044330-1, e-ISBN 978-3-11-044331-8

Downstream Processing in Biotechnology

Edited by
Venko N. Beschkov and Dragomir Yankov

DE GRUYTER

Editors
Prof. Dr. Venko N. Beschkov
Bulgarian Academy of Sciences
Institute of Chemical Engineering
akad.G.Bonchev str. bl.103
1113 Sofia
Bulgaria

Prof. Dr. Dragomir Yankov
Bulgarian Academy of Sciences
Institute of Chemical Engineering
akad.G.Bonchev str. bl.103
1113 Sofia
Bulgaria

ISBN 978-3-11-057395-4
e-ISBN (PDF) 978-3-11-057411-1
e-ISBN (EPUB) 978-3-11-057400-5

Library of Congress Control Number: 2021935349

Bibliographic information published by the Deutsche Nationalbibliothek
The Deutsche Nationalbibliothek lists this publication in the Deutsche Nationalbibliografie; detailed bibliographic data are available on the Internet at http://dnb.dnb.de.

Cover image: Stefano Gilera/Cultura/Getty Images
Typesetting: Integra Software Services Pvt. Ltd.
Printing and binding: CPI books GmbH, Leck

www.degruyter.com

Preface

Traditional biotechnologies are well known from ancient times. Alcohol beverages, dairy products, and other food products are used for many centuries throughout the world. The 20th century brought many innovative technologies driven by the demand for energy efficiency, environment protection, health issues, etc. Biotechnologies played a very important role in this direction. Their main advantage of fermentation processes consists of the apparently "one step" processes compared to the processes in the traditional chemical technologies. Further, the microbial processes are selective and they produce hiral products with a high biological activity which is hardly possible for chemical methods.

On the other hand there are certain disadvantages of fermentation processes. They are accomplished by living microbial cells and therefore they run with very low conversion rates compared to the chemical processes. Usually, the microbial cells are sensitive to substrate and product concentrations with considerable inhibition. That is why the final product concentrations are relatively low. Moreover, these products are frequently unstable at higher temperatures. All of these features lead to difficulties in product isolation and purification. The complex processes associated with product extraction, concentration, isolation, and purification are known as "downstream processing".

As a final step in fermentation technologies, downstream processing is a very important part of industrial biotechnology following the fermentation. Downstream processing is so important for the whole process of product formation by biotechnologies because as a rule, the concentrations of the target products are usually very low. This fact implies long, tedious, and energy-consuming processes which may compromise the very biotechnology. According to some authors, the expenses for downstream processing may reach 50% of the total costs for certain product manufacturing.

The large scale processes for product concentration and isolation are mostly associated with similar processes used in chemical technology: sedimentation, filtration, extraction, evaporation, distillation, drying. These processes are applied mostly for large scale manufacturing. However, the specific properties of many biotechnological products as well as their low concentrations in the final broth need some more sophisticated approach for product isolation, purification, and recovery. That is why new approaches for product recovery are desired.

The main difference between separation processes in chemical technology and biotechnology is the presence of the living cells in the mixture to be separated. In the case of intracellular or membrane-bound products, these steps are preceded by a cell's disruption step.

When the target product is in the fermentation broth, further product extraction is preceded by evaporation to remove the excessive amount of water to concentrate the desired products to an acceptable concentration. When the product concentrations in

https://doi.org/10.1515/9783110574111-202

the broth are usually low a huge amount of water must be evaporated to attain feasible concentrations being appropriate for further processing. That is why membrane processes are proposed to avoid product and heat losses during traditional evaporation.

Extraction is the most spread technique for selective isolation of fermentation products. It is based on the different solubility of the products in water and organic solvents and enables to invent and apply various compositions of the active extracting agent tailored for each specific case. Combinations of extraction and membrane processes reveal many powerful systems for product isolation with high yield.

Application of aqueous two-phase systems for biomass separation is a method that has no equivalence in chemical technology.

When the fermentation products are ionogenic, ion exchange may be a suitable method for product isolation and purification to avoid energy spending for evaporation.

The present book presents a review of the most used methods in downstream processing in biotechnology. Those are extraction in different ways: liquid/liquid extraction, supercritical one, membrane processes, product separation in two-phase aqueous systems and ionic liquids, and ion exchange techniques.

There is another broad area for product separation and isolation as chromatography. Many of the fermentation products, like biologically active proteins, vaccines, etc. after crude separation are extracted and purified by a variety of chromatographic methods. We hope it deserves specific attention by other authors in other issues dedicated specifically to it.

Venko N. Beschkov
Dragomir S. Yankov
editors

Contents

List of Contributing Authors

Katalin Belafi-Bako
Research Institute on Biochemical Engineering
Membrane Technology and Energetics
University of Pannonia
Egyetem u. 10., 8200 Veszprem, Hungary
E-mail: bako@almos.uni-pannon.hu

Venko Beschkov
Institute of Chemical Engineering
Bulgarian Academy of Sciences
Acad.G.Bonchev St.
Block 103, Sofia, 1113 Bulgaria
E-mail: vbeschkov@yahoo.com

José Coelho
Instituto Politecnico de Lisboa, Instituto Superior de Engenharia de Lisboa, CIEQB and DEQ, Rua Conselheiro Emidio Navarro 1, 1959-007 Lisboa, Portugal
Instituto Superior Técnico, Centro de Quimica Estrutural, Universidade de Lisboa, Av. Rovisco Pais, 1 1049-001 Lisboa, Portugal
E-mail: jcoelho@deq.isel.ipl.pt

Konstantza Tonova
Institute of Chemical Engineering
Bulgarian Academy of Sciences
Acad. G. Bonchev Str., Bldg. 103
1113, Sofia, Bulgaria
E-mail: konstantzatonova@yahoo.com; k.tonova@iche.bas.bg

Svetlozar Velizarov
3 LAQV, Chemistry Dept./FCT
Universidade Nova de Lisboa
2829-516 Caparica, Portugal
E-mail: s.velizarov@fct.unl.pt

Kailas Wasewar
Advance Separation and Analytical Laboratory (ASAL)
Department of Chemical Engineering
Visvesvaraya National Institute of Technology (VNIT)
Nagpur, 440010 India
E-mail: k_wasewar@rediffmail.com

Dragomir Yankov
Institute of Chemical Engineering
Bulgarian Academy of Science
Acad. G. Bontchev str, block 103
Sofia 1113, Bulgaria
E-mail: dsyankov@gmail.com

https://doi.org/10.1515/9783110574111-204

V. Beschkov and D. Yankov

1 Chemical engineering methods in downstream processing in biotechnology

Abstract: Downstream processing in industrial biotechnology is a very important part of the overall bioproduct manufacturing. Sometimes the cost for this part of biotechnologies is up to 50% of the overall expenses. It comprises product concentration, separation and purification to different extents, as requested. The usually low product concentrations, the large volumes of fermentation broth and the product sensitivity toward higher temperatures lead to specific methods, similar but not identical to the ones in traditional chemical technology.

This article summarizes briefly the unit operations in downstream processing in industrial biotechnology, making a parallel between biotechnology and chemical technology.

Keywords: biotechnology, product recovery, chemical engineering methods

1.1 Introduction

Chemical engineering appeared in the beginning of the XXth century when the oil industry started to play important role in fuel production. Later oil had become an important source for different industrial chemical synthesis, such as plastic production, fine chemical syntheses, detergents, etc. Chemical engineering appeared to be very important for chemical technologies either, enabling the selection of more appropriate equipment and processes for better yields of high quality products.

The main processes met in chemical technologies are mass transfer in multiphase media (gas absorption, liquid/liquid and solid/liquid extraction, solid dissolution, adsorption), heat transfer (at distillation, evaporation, drying), chemical reactions in homogeneous and heterogeneous systems, catalytic conversions, etc. They can be considered separately in each particular case depending on the product properties, biomass peculiarities, etc. Another approach is the optimization of entire systems for chemical technologies accomplishment selecting the optimum process interaction in different apparatuses.

Industrial biotechnology takes the advantages and the experience of chemical technologies using similar (or exactly the same) processes and similar equipment. That is why chemical engineering science is of a great help in biotechnology design,

This article has previously been published in the journal Physical Sciences Reviews. Please cite as: Beschkov, V., Yankov, D. Chemical engineering methods in downstream processing in biotechnology *Physical Sciences Reviews* [Online] 2021, 1. DOI: 10.1515/psr-2018-0064

https://doi.org/10.1515/9783110574111-001

operation and optimization. However, the operation with microbes in biotechnologies poses some differences compared to chemical technologies. The main difference between separation processes in chemical technology and biotechnology is the presence of the living cells in the mixture to be separated. Another challenge is the bioactivity of target products, which must be preserved. Sometimes the substance of interest is bound to the cell's membrane and cells must be disrupted.

A comparison of these two types of industrial technologies is shown in Table 1.1.

Table 1.1: Comparison of chemical technologies and biotechnologies.

Item	Chemical technologies	Biotechnologies
Mode of operation	Multi-step technology, many processes of intermediate isolation; batch and continuous processes	Fermentation: apparently single step final product production; batch processes mostly
Operating conditions	High temperatures and pressures; high-cost equipment	Ambient temperatures and pressures; low-cost equipment
Catalysts	Catalyst recovery required	Catalysts produced during the fermentation process
Reaction selectivity	Low reaction selectivity; racemic mixtures only	High hiral product selectivity
Operation time	Fast reactions; short time processes	Slow processes; long duration; energy consumption
Sterilization	No sterilization required	Sterilization required
Product inhibition	Variable	Strong product inhibition; low product concentrations

As a final step, downstream processing is a very important part of industrial biotechnology following the fermentation (being the key-process) and associated with isolation of target products and their concentration and purification. According to some authors, the expenses for downstream processing may reach 50% of the total costs for certain product manufacturing [1].

Downstream processing is so important for the whole process of product formation by biotechnologies because as a rule the concentrations of the target products are usually very low.

This fact implies long, tedious and energy consuming processes which may compromise the very biotechnology.

The large-scale processes for product concentration and isolation are mostly associated with similar processes used in chemical technology: sedimentation, filtration, extraction, evaporation, distillation, drying. These processes are applied mostly for large-scale manufacturing. However, the specific properties of many biotechnological products as well as their low concentrations in the final broth need some more sophisticated approach for the product isolation, purification and recovery. That is why, new approaches for product recovery are desired.

Typical bioseparation processes usually include the following steps – cell separation, product concentration, primary purification, and final purification and polishing, cf. Figure 1.1. In the case of intracellular or membrane-bound products, these steps are preceded by a cell's disruption step. Various separation technics can be used at each step.

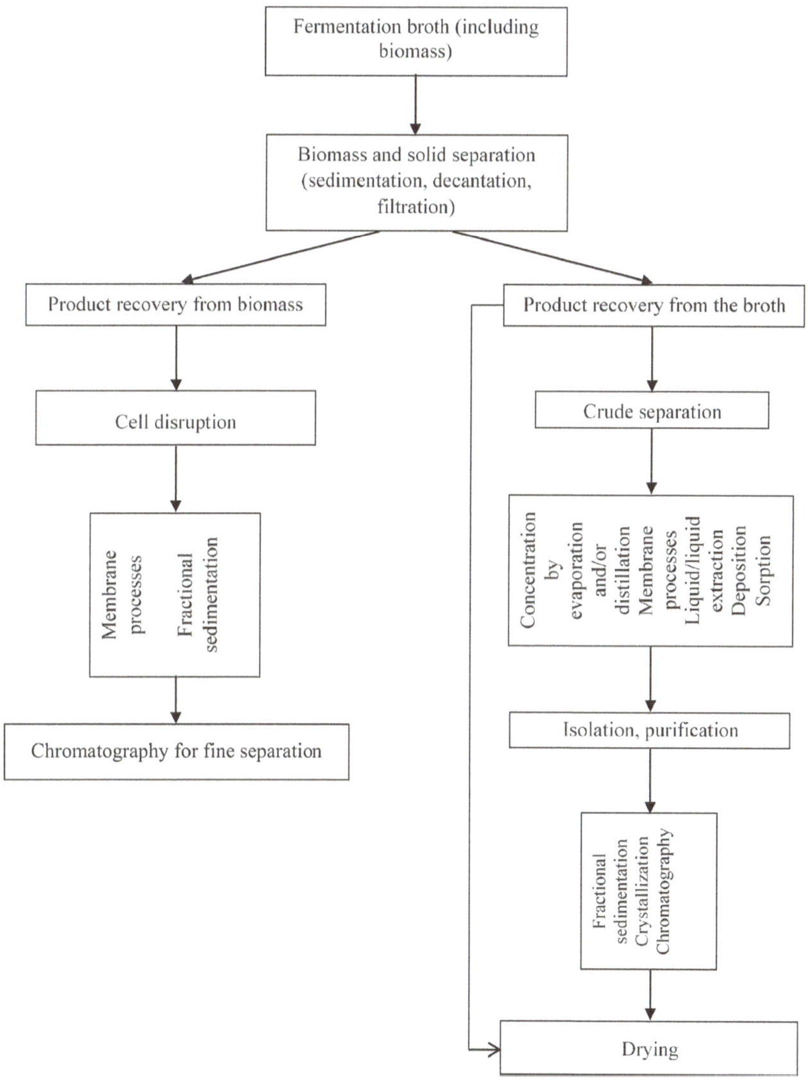

Figure 1.1: Principal flow sheet of downstream processing in industrial biotechnology.

Several factors must be considered when designing a bioseparation process – type and nature of the starting material, the location of the target substance, volume and concentration of the starting material, stability and bioactivity of the target product, the final physical form of the product, purity requirements, waste minimization, and total process cost.

This article proposes comparative analysis of the traditional methods of chemical technology and their improvements and alternatives for product recovery in industrial biotechnology.

1.2 Main and specific processes for product extraction and recovery in biotechnology

There are two main paths or approaches in downstream processing (cf. Figure 1.1). They are associated with the target product location: within the microbial cells (intra-cellular) or extra-cellular, i.e. in the broth. In both cases separation of the biomass from the liquid phase is required.

1.2.1 Biomass separation

First, biomass must be concentrated and afterwards removed by filtration. In the case of bacterial biomass its concentration is hampered by its density comparable to the one of the aqueous broth. That is why, the biomass concentration should be made by enhanced sedimentation by adding coagulants and flocculants to the broth.

Coagulants neutralize the electrical charge on colloid particles and decrease their z-potential, which reduces the forces keeping colloid particles apart and facilitates their agglomeration. In the cases of wastewater treatment, salts of poly-valent cations (Al^{3+}, Fe^{3+}, etc.) are used. According to the well-known Schulze-Hardy rule the higher the valence of the added ions, the lower the stability of the colloid system. Coagulants are practically used for removal of fine suspensions in waste water and for floating biomass separation.

Flocculants are used for enhanced solid sedimentation and can be applied for biomass separation too. They are electrically charged polymers: poly-sulfonates, polycarbonates (when negatively charged) or poly-amines, when positively charged. One of these types is selected depending on the charge of the colloid particles. The latter are attracted by the oppositely charged polymer molecules and therefore agglomerated and easily deposited.

Filamentous microbes have higher surface area and volume compared to conventional flocs, also they are slow to settle. Under particular conditions, these microbes can grow excessively, which will cause issues with sludge settling thereby decreasing the efficiency of the wastewater treatment plant [2]. Wide-ranging research is continuing to be able to establish what causes a growth of a certain type of filamentous microorganisms. It was observed that season changes and difference in ambient temperature could be one of the reasons of the plentiful growth of filamentous microbes.

After coagulation and flocculation the biomass-free broth is decanted and the enriched sediment with considerably lower volume and water content is subjected to filtration.

Filtration is accomplished by machines well known in chemical technology and wastewater treatment [2]. They are filter presses, rotating drum vacuum filters or centrifuges [3–5]. In all cases, the major problem is the filtration cake fouling the filter pores making the process slow, long lasting and inefficient. For this reason the concentrated biomass is mixed with by structure forming expanded sand forming a porous structure, permeable for the filtrate.

After washing, the cake could be subjected to further processing if a valuable product is contained, or deposited. Sometimes it has been used as animal feed additive. In case the target product is in the broth its recovery takes place, according to a downstream processing scheme with processes shown in Figure 1.1.

Application of aqueous two-phase systems (ATPS). Application of aqueous two-phase systems for biomass separation is a method that has no equivalence in chemical technology. ATPS are spontaneously formed when two water-soluble polymers are mixed in a common solution over a certain concentration. Because of the formation of large polymer-polymer aggregates and steric effects, the systems separate in two distinct water phases – one rich of the first polymer and the other rich of the second. ATPS are characterized by low interfacial tension, low viscosity, fast phase's separation, high separation yield, high biocompatibility, and effortless scale-up, among others. ATPS can be also formed by mixing of a polymer and salts like phosphates, sulfates, citrates, etc., low molecular aliphatic alcohols and salts. Recently surfactants-based as well as ATPS formed using ionic liquids and deep eutectic solvents were investigated. Properties of ATPS can be easily manipulated by varying polymers and salts concentrations, temperature, pH, polymer's molecular weight and salt's nature, thus governing the partition of the desired product. As a rule, large objects (cells, cells debris, and organelles) distribute unevenly in one of the phases what make ATPS an ideal choice for a primary step in product removal from the fermentation broth. Regardless of the attractive properties of ATPS for separation of biotechnology products, their application in industry is still hampered by problems with recycling of phase components and environmental problems. Application of ATPS in large scale is reviewed by Torres-Acosta et al. [6]

One can see that the methods for biomass separation do not differ considerably from the ones used in chemical technology and in traditional wastewater treatment. For certain specific cases a novel techniques (ATPS) has promising features.

1.2.2 Target product extraction from disrupted biomass

Although nearly all products of microbial origin are excreted into the fermentation broth a large proportion of the potentially useful microbial products is retained within the microbial cells. Those products are various: many enzymes, e.g. L-asparaginase, catalase, cholesterol oxidase, glucose isomerase [7], genetic material [8], etc. The isolation of intracellular material requires that the cell must be disintegrated by physical, chemical or enzyme methods to release its contents into the surrounding medium [7]. However, microorganisms are more robust than it is generally expected. The resistance to disruption of microorganisms is due to their internal osmotic pressure [9]. There are different methods for cell disruption, but some of them are of lab-scale importance only, i. e. enzyme methods [10], or ultrasonication [11, 12].

More practical applications deserve the mechanical methods – grinding in bead mills [7], disruption under high pneumatic or hydraulic pressure (high pressure homogenization) [13, 14] and microfluidization [15]. However, it was pointed out that mechanical methods are highly energy consuming [16]. Another group of methods are based on the sudden change of physical conditions the cells are exposed to. For example, two of them are nitrogen decompression [17] and cryo-pulverization [15].

One well known practice is the ultrasonic disintegration [7, 10, 11, 18, 19]. It is based on the induction of high shear rates in the liquid and cavitation leading to cell disruption and release of the content into the broth. Because of its price, adverse impact on living beings and scale limitations it is recommended mainly for lab-scale purposes or small scale industrial applications. Another hinder to ultrasonic disintegration in bioprocessing is the released heat leading to damages in the target products, being mostly proteins. That is why intensive cooling is required. Ultrasonication may lead to chemical changes in molecules, like formation of free radicals in the broth [19]. Nevertheless, there are some efforts for larger scale practical applications, e.g. in lipids recovery from microalgae [20].

The further processes of product extraction are based on cell debris removal by filtration (combined with centrifuging), sometimes necessary for target enzyme recovery from the debris. The next step is to process the liquid phase to extract the valuable products. As a rule these products are mostly proteins and the applied extraction methods are typical for protein engineering [21, 22], membrane methods, like ultrafiltration [23–25], etc.

One of the mostly spread methods for protein processing and separation is fractional sedimentation by change of the liquid media composition (adding ethanol or acetone, addition of inorganic salts, pH changes), thus separating the proteins by their solubility in the new medium [26, 27]. After this crude separation the target proteins are finely separated by gel permeation chromatography [28], membrane [29] and affinity chromatography [30]. The latter methods are mostly applicable for small scale products with very high added value.

There are some new efforts for lipid extraction from microalgae for biodiesel production [16, 21]. This group of methods is associated with the properties of microbial cells which have no analogue in chemical technology.

1.2.3 Product recovery from the broth. Crude separation and concentration

Crude separation. After removal of coarse particles, solids and biomass as it was mentioned above, the next step is product concentration by different techniques. The most spread one is evaporation, accompanied by distillation.

Evaporation and distillation. The purpose of evaporation is to remove the excessive amount of water to concentrate the desired products to acceptable concentration. Then the consequent processes for product separation are admissible. Since the product concentrations in the broth are usually low (i.e. up to 10 g dm^{-3}) it is evident that huge amount of water must be evaporated to attain concentrations about 100 g dm^{-3} being appropriate for further processing. In cases when the next step is crystallization the situation becomes worse. Taking into account that the majority of fermentation products are thermally unstable one can conclude that evaporation must be accomplished at low temperatures, i.e. under vacuum.

Other high-speed evaporation techniques are based on thin film evaporation either under atmospheric conditions or in vacuum [31, 32]. One of these approaches is to spread the liquid solution onto a heated rotating disc, thus forming a thin film on it. Because of the small film thickness (less than 0.1 mm) the heat transfer rate is sufficiently high to heat the liquid and to attain very high evaporation rates, thus avoiding long contact times and keeping the product quality. In these cases the heat transfer coefficients reach 8 kW $m^{-2}K^{-1}$. The high heat transfer rate enables to operate at medium temperatures (i.e. about 50° C).

Some other equipment based on the thin liquid film evaporation has been proposed and applied. In their construction the evaporated liquid is spread onto the inner surface of cylinder heated from outside [33]. The solution is spread on the inner cylindrical surface by slowly rotating blade (wiped film evaporator). Such apparatuses have been produced by the world well known company α-Laval, for example, see [34]. The same company has proposed a rising film evaporator with up- and down-directed flow along the heat exchanger [35].

There are also thin film evaporators based on falling films on a free vertical smooth surface, usually cylindrical, for example, SMS production [36]. A principal sketch of such an evaporator is shown in Figure 1.2.

All of the thin film evaporators have been applied for decades. All of these methods and equipment are successfully applied in chemical and food industry by various applications [37].

Distillation is another process developed for chemical and oil industries but it has been successfully applied in biotechnology. The classical approach in

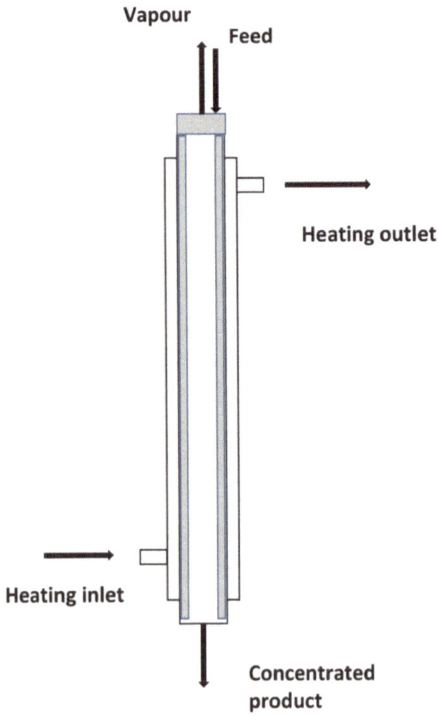

Figure 1.2: Falling film evaporator.

chemical technology, i.e. the vacuum distillation, is typical for biotechnological purposes, because of the thermal sensitivity of many bio-products. It still has serious application, in ethanol production, for example. In these cases distillation columns for 95.8% vol. ethanol production (or other desired concentrations) are applied [38].

It is well known that ethanol concentration cannot reach more than 100 g dm^{-3} during fermentation, because of the product inhibition. One possible and attractive option is to couple the fermentor with distillation unit to remove continuously the ethanol within and hence to shift the fermentation process to complete substrate utilization [39].

Evaporation and distillation are well known operations traditionally exploited in chemical technology where heat and mass transfer processes are in the basis of design and practical applications. There are very strong similarities for the use of these processes in chemical technology, heat engineering and industrial biotechnology.

Membrane processes. At very low product concentrations when evaporation is not feasible membrane processes are recommended. They are ultrafiltration, reverse osmosis and electrodialysis. Various membrane separation methods have been applied depending on the particle size to be removed or recovered (cf. Table 1.2).

Table 1.2: Particle size range and different membrane separation techniques.

Particle size	1 nm	10 nm	100 nm	1 µm	10 µm	0,1 mm	1 mm
Substance to separate	Dissolved salts, sugars	Viruses	Bacteria		Yeasts	Coarse particles	
Separating process	Reverse osmosis	Ultrafiltration				Particle filtration	
		Nanofiltration	Microfiltration				

Membrane processes are increasingly used in pharmaceutical and biochemical engineering and biotechnology for concentration and purification, for synthesis of molecules and drug delivery systems, and support for biochemical reactions [40, 41]. Separation and concentration by ultrafiltration are based on passing the liquid medium across a membrane with calibrated pore size, thus enabling to retain certain molecules [42–45]. Ultrafiltration membranes typically separate compounds with molecular weights from 1,000 to 1,000,000 daltons (i.e. 1 to 100 nm in size). Besides ultrafiltration [42], microfiltration [40], membrane chromatography [28], membrane emulsification, liquid membranes and membrane bioreactors are also studied and applied.

Reverse osmosis is another membrane process based on applying pressure on the diluted solution or broth to overcome the osmotic pressure and to provoke trickling of the solvent through the membrane and thus to concentrate the primary solution [43–45]. This method is extensively applied for sea water desalination [43], removal of organic pollutants [44] and it has been proposed for product concentration from fermentation broth, like lactic acid fermentation [45].

Membrane processes are considerably applied in water desalination, protein concentration and separation with their specific applications in biotechnology where low concentrations of thermally unstable products are frequently met. The common issue for chemical technology and biotechnology is the modelling and design for these membrane processes.

Liquid/liquid extraction. After crude concentration by vacuum evaporation or membrane processes, more refined extraction of the target product is the next step. In case of organic products with a good solubility in organic solvents, like antibiotics, solvent extraction is appropriate method to extract selectively and further to recover the product with a very good purity after solvent distillation.

The selection of solvent depends on different requirements and constraints. First, it must not be miscible with water in order to avoid water pollution and to increase extraction efficiency. Next, solvents must be less flammable and explosively safe. These two conditions are scarcely satisfied, but organic solvents with high boiling points and high temperature of explosion are preferred. Such solvents for instance are *n*-paraffins, butyl acetate, etc.

There are three main extraction types of facilities known from chemical industry: the mixer-settler, operating under batch conditions, counter- or co-current extraction columns and centrifugal extraction equipment [46].

The first one is not suitable for large-scale production because of its low productivity, and because of the difficulties associated with stable emulsion formation and very slow phase separation.

The most appropriate extraction technique in biotechnology is centrifugal extraction. There are two types of centrifugal equipment: mixer and settler operation in centrifugal field or combined ones. The merits of this process are better pronounced when combined mode of operation is applied. Then fine dispersion and quick phase separation in the rotating drum of the centrifuge take place. This type of devices operate both under batch or continuous conditions in very intensive mode.

In the considered case the denser phase is introduced close to the axe of rotation and the light one – to the drum periphery. Afterwards, the natural centrifugal force tends to direct the two phase to each other hence to facilitate their mixing and further separation.

In some cases when the target product is ionic, ion exchange extraction is applied [47]. In such cases, when organic acids are the target product, the organic phase contains active component (e.g. amine) to form quaternary ammonia salt, soluble in the organic solvent. Afterwards (or simultaneously) the target anion is stripped by aqueous solution of inorganic base to recover the product and to regenerate the solvent.

Liquid/liquid extraction is frequently applied for organic products, like antibiotics, lipid-soluble vitamins [48]. Once the biomass is removed, there are practically no differences in the extraction processes in chemical technology and biotechnology. That is why the methods for modelling and design are quite similar.

1.2.4 Crystallization and drying

The solid product separation is accomplished by crystallization by cooling of the concentrated solution or by evaporation drying in a flow of hot air.

Crystallization by cooling is operating in batch or continuous mode. Afterwards the crystals are washed by cold water (when water-soluble) and recovered by centrifuging. There is a method of coupled concentration by evaporation and crystallization (so-called Oslo-crystallizer [49]) where the product-containing solution is passed through heat exchanger for solvent evaporation with consequent cooling in another unit where crystallization takes place.

There are various types of crystallizers to be chosen depending on the specific properties and the requirements for the products. These issues are discussed in [50–52].

There are some new approaches for enantiomer separation [53] and enzyme crystallization by membrane technique [54].

Drying is essential process for the final product preparation. It is applied either for already purified and concentrated product containing solutions, or for direct obtaining of product rich powder concentrate. The latter is adopted for low-grade preparations, like L-lysine containing additives for animal feed, single cell proteins, etc. However, attention must be paid on the thermal sensitivity of the products, either microbial cells or chemicals [55].

There are various design of dryers: operating in jet-spouted bed, spray, fluidized bed, etc., cf. [55]. Freezing and vacuum drying are available as well.

Lyophilization (or freeze drying) is another option to dry thermally unstable products [56, 57]. It consists in cooling the treated solution at high pressures to very low temperatures (under the triple point in the water phase diagram) and after lowering the pressure to provoke sublimation of water and drying. Besides the application in food industry, many small scale products of biotechnology are produced in this way, e. g. vaccines [57], some fine pharmaceuticals [58], bio-active proteins and enzymes [59], various kinds of bacteria; etc.

In general, besides lyophilization, as design and process operation crystallization and drying are quite similar to the processes in chemical industry.

1.2.5 Emerging bioseparation processes

The developments in the production of novel biotechnology products like monoclonal antibodies, plasmid DNA, recombinant RNA, human growth hormone, insulin-like growth factor or virus-like particles leads to the emerging or development of new separation techniques integrating into one, several separations and purification steps.

- Application of aqueous two-phase systems (ATPS) – new system types, affinity systems, thermosensitive phase-forming constituents etc.
- Chromatography – affinity chromatography (AC), size-exclusion chromatography (SEC), hydrophobic-interaction chromatography (HIC), ion-exchange chromatography (IEX) – new stationary phases, increasing of the selectivity and capacity, different materials for monolith columns.
- Expanded bed adsorption, involving new sorbents.
- Fibre based adsorption systems.
- Convective flow systems and devices.

1.2.6 Separation by chromatography

This group of methods are typical and in many cases inevitable for separation and recovery of protein products from various manufacturing. This issue comes from the specific properties of these products, like thermal instability, chemical and pH sensitivity and the loss of biological activity under extreme conditions. That is why, the traditional methods of chemical technology, like evaporation, deposition by chemicals, etc. are inadmissible. That is why, after crude separation by ultrafiltration and fractional sedimentation more refined chromatographic methods are required. However, this is quite broad topic, which is beyond the purpose of this overview.

1.3 Conclusions

The purpose of this review is to make a brief survey of the downstream processing in industrial biotechnology and to outline the similarities between its methods and the processes in chemical technology, where the tools of chemical engineering are applied.

It is evident that besides the similar techniques and equipment, there are big differences, resulting on the specificities of some products in biotechnology and the ones from chemical industry.

Therefore, selective approach based on the product properties and the operational costs must be adopted.

References

[1] Bishai M, De S, Adhikari B, Banerjee R. A platform technology of recovery of lactic acid from a fermentation broth of novel substrate *Zizyphus oenophlia*. 3 Biotech. 2015;5:455–63. DOI: [10.1007/s13205-014-0240-y].

[2] Tchobanoglous G, Burton FL, Stensel HD. Wastewater engineering: treatment and reuse. 4th ed. Mumbai: Tata McGraw-Hill Education, 2011.

[3] Perlmutter BA. Improving process operations with a rotary pressure filter. BHS-Filtration Inc., Date of retrieve: 31 sep 2013, [Online]. 2000. http://www.bhs-filtration.com/improvingProcOpsRotary.pdf.

[4] BOKELA Rotary Drum Filters [Online]. http://www.bokela.de/uploads/media/TFI-prosp_e_06.p.

[5] Wiesmann U, Binder H. Biomass separation from liquids by sedimentation and centrifugation. In: Reaction engineering. Springer Berlin Heidelberg, Jan 1970:119–71. DOI:10.1007/3-540-11699-0_12.

[6] Torres-Acosta MA, Mayolo-Deloisa K, González-Valdez JE, Rito-Palomares M. Aqueous two-phase systems at large scale: challenges and opportunities. Biotechnol J. 2019;14:1800117. DOI:10.1002/biot.201800117.

[7] Chisti Y, Moo-Young M. Disruption of microbial cells for intracellular products. Enzyme Microb Technol. 1986;8:194–204.

[8] Wimpenny JW. Breakage of microorganisms. Process Biochem. 1967;2:41–4.

[9] Tam YJ, Allaudin ZN, Lila MA, Bahaman AR, Tan JS, Rezaei MA. Enhanced cell disruption strategy in the release of recombinant hepatitis B surface antigen from *Pichia pastoris* using response surface methodology. BMC Biotechnol. 2012;12:70.

[10] Andrews BA, Asenjo JA. Enzymatic lysis and disruption of microbial cells. Trends Biotechnol. 1980;5:273–7.
[11] Tangtua J. Evaluation and comparison of microbial cells disruption methods for extraction of pyruvate decarboxylase. Int Food Res J. 2014;21:1331–6.
[12] Gerde A, Montalbo-Lomboy M, Yao L, Grewell D, Wang T. Evaluation of microalgae cell disruption by ultrasonic treatment. Bioresour Technol. 2012;125:175–81.
[13] Wang DI, Cooney CL, Demain AL, Dunnill P, Humphrey AE, Lilly MD. Fermentation and enzyme technology. New York: John Wiley, 1979:238.
[14] Howlader MS, French WT, Shields-Menard SA, Amirsadeghi M, Green M, Rai N. Microbial cell disruption for improving lipid recovery using pressurized CO_2: role of CO_2 solubility in cell suspension, sugar broth, and spent media. Biotechnol Prog. 2017;33:737–48. DOI:10.1002/btpr.2471.
[15] Choi H, Laleye L, Amantea GF, Simard RE. Release of aminopeptidase from *Lactobacillus casei* sp. casei by cell disruption in a microfluidizer. Biotechnology Techniques. 1997;11:451–3.
[16] Khot M, Ghosh D. Lipids of *Rhodotorula mucilaginosa* IIPL32 with biodiesel potential: oil yield, fatty acid profile, fuel properties. J Basic Microbiol. 2017. DOI:https://doi.org/10.1002/jobm.201600618.
[17] Lee AK, Lewis DM, Ashman PJ. Disruption of microalgal cells for the extraction of lipids for biofuels: processes and specific energy requirements. Biomass Bioenergy. 2012;46:89–101.
[18] Brown MR, Sullivan PG, Dorenbos KA, Modafferi EA, Geddes JW, Steward O. Nitrogen disruption of synaptoneurosomes: an alternative method to isolate brain mitochondria. J Neurosci Methods. 2004;137:299–303.
[19] Suslick KS. 1998. Kirk-Othmer encyclopedia of chemical technology. 4th ed. New York: J. Wiley & Sons, Vol. 26, 1998:517–41.
[20] Jeon B-H, Choi J-A, Kim H-C, Hwang J-H, Abou-Shanab RA, Dempsey BA, et al. Ultrasonic disintegration of microalgal biomass and consequent improvement of bioaccessibility/bioavailability in microbial fermentation. Biotechnol Biofuels. 2013;6:37. DOI:10.1186/1754-6834-6-37.
[21] Lee A, Lewis D, Ashman P. Disruption of microalgal cells for the extraction of lipids for biofuels: processes and specific energy requirements. Biomass Bioenergy. 2012;46:89–101.
[22] Wardhan R, Mudgal P. 2017. Textbook of membrane biology. Singapore: Springer. pp. 49–60. DOI:org/10.1007/978-981-10-7101-0_3.
[23] Moore SM, Hess SM, Jorgenson JW. Extraction, enrichment, solubilization, and digestion techniques for membrane proteomics. J Proteome Res. 2016;15:1243–52.
[24] Cheang B, Zydney AL. A two-stage ultrafiltration process for fractionation of whey protein isolate. J Membrane Sci. 2004;231:159–67.
[25] Tutunjian RS. Ultrafiltration processes in biotechnology. Ann New York Acad Sci. 1983;413:238–53.
[26] Martin RG, Ames BN. A method for determining the sedimentation behavior of enzymes: application to protein mixtures. The J Biol Chem. 1961;236:1372–9.
[27] Englard S, Seifter S. Precipitation techniques. Methods Enzymol. 1990;182:285–300.
[28] Hashimoto T, Sasaki H, Aiura M, Kato Y. High-speed aqueous gel-permeation chromatography. J Polymer Sci Polymer Phys. 1978;16:1789–800.
[29] Zeng X, Ruckenstein E. Membrane chromatography: preparation and applications to protein separation. Biotechnol Prog. 1999;15:1003–19.
[30] Cuatrecasas P, Wilchek M, Anfinsen CB. Selective enzyme purification by affinity chromatography. Proc National Acad Sci U S A. 1968;61:636–43.
[31] Thin film evaporator. Patent No.: US 5. 256, 250. 26 Oct 1993.

[32] Thin-film evaporator. Patent No.: US 7. 591, 930. 22 Sep 2009.
[33] https://lcicorp.com/thin_film_evaporation/thin_film_wiped_film_evaporator.
[34] http://www.aaronequipment.com/usedequipment/evaporators/wipe-film-thin-film/alfa-laval-ct-6-47676001.
[35] https://www.alfalaval.com/products/heat-transfer/plate-heat-exchangers/gasketed-plate-and-frame-heat-exchangers/alfavap/.
[36] http://www.360evaporator.com/falling-film-evaporator.html.
[37] https://www.sms-vt.com/technologies/drying-technology/vertical-thin-film-dryer/?gclid=EAIaIQobChMIjb6389ea3wIVl5IYCh0w5wg_EAAYAiAAEgIKl_D_BwE.
[38] Katzen R, Madson PW, Moon GD Jr. Ethanol distillation: the fundamentals, CHEE332, Queens University, Kingston, Ont., Chapter 18. 269–88. https://chemeng.queensu.ca/courses/CHEE332/files/distillation.pdf.
[39] Cutzu R, Bardi L. Production of bioethanol from agricultural wastes using residual thermal energy of a cogeneration plant in the distillation phase. Fermentation. 2017;3:24. DOI:10.3390/fermentation3020024.
[40] Nelson BK, Barbano DM. A microfiltration process to maximize removal of serum proteins from skim milk before cheese making. J Dairy Sci. 2005;88:1891–900.
[41] Ghosh R. Protein separation using membrane chromatography: opportunities and challenges. J Chromatogr A. 2002;952:13–27.
[42] O'Sullivan TJ, Beaton NC, Epstein AC, Korchin SR. Applications of ultrafiltration in biotechnology. Chem Eng Prog. 1984;80:68–75.
[43] Jamaly S, Darwish NN, Ahmed I, Hasan SW. A short review on reverse osmosis pretreatment technologies. Desalination. 2014;354:30–8.
[44] Williams ME, Hestekin JA, Smothers CN, Bhattacharyya D. Separation of organic pollutants by reverse osmosis and nanofiltration membranes: mathematical models and experimental verification. Ind Eng Chem Res. 1999;38:3683–95.
[45] Schlicher LR, Cheryan M. Reverse osmosis of lactic acid fermentation broths. J Chem Technol Biotechnol. 1990;49. DOI:https://doi.org/10.1002/jctb.280490204.
[46] Schuegerl K. Solvent extraction in biotechnology, recovery of primary and secondary metabolites. Springer Berlin Heidelberg, 1994.
[47] Yordanov B, Boyadzhiev L. Pertraction of citric acid by means of emulsion liquid membranes. J Membr Sci. 2004;238:191–7.
[48] Vieira Dos Santos N, de Carvalho Santos-ebinuma V, Pessoa Junior A, Brandão Pereira JF. Liquid–liquid extraction of biopharmaceuticals from fermented broth: trends and future prospects. J Chem Technol Biotechnol. 2017;13. DOI:https://doi.org/10.1002/jctb.5476.
[49] Mullin JW, Nyvlt J. Design *of* classifying crystalisers. Trans Instn Chem Engrs. 1970;48:7–14.
[50] Alhalabi T, Koikkalainen K, Ern LS. CHEM-3140 – Bioprocess technology II. Drying and crystallization. Aalto University, School of Chemical Technology, Espoo, Finland, 2017.
[51] Kent JA. editor. Kent and Riegel's handbook of industrial chemistry and biotechnology. Springer US, 2010:1686.
[52] Kristiansen B, Linden J, Mattey M. Citric acid biotechnology. Philadelphia: Taylor&Francis Inc. Pp. 182, 183. 2002.
[53] Majumder A, Nagy ZK. A comparative study of coupled preferential crystallizers for the efficient resolution of conglomerate-forming enantiomers. Pharmaceutics. 2017;9:55. DOI:10.3390/pharmaceutics9040055.
[54] Di Profio G, Perrone G, Curcio E, Cassetta A, Lamba D, Drioli E. Preparation of enzyme crystals with tunable morphology in membrane crystallizers. Ind Eng Chem Res. 2005;44:10005–12. DOI:10.1021/ie0508233.

[55] Adamiec J, Kaminski W, Markowski AS, Strumiłło C. Ch. 39, Drying of biotechnological products. In: Handbook of industrial drying, (A.S. Mujumdar, editor). Bosa Roca USA. Taylor&Francis Inc., 2006:906–29.
[56] Nireesha GR, Divya L, Sowmya C, Venkateshan N, Niranjan Babu M, Lavakumar V. Lyophilization/freeze drying - an review. Int J Novel Trends Pharmac Sci. 2013;3:87–98.
[57] Hansen LJ, Daoussi R, Vervaet C, Remon JP, De Beer TR. Freeze-drying of live virus vaccines: A review. Vaccine. 2015;33:5507–19. DOI:10.1016/j.vaccine.2015.08.085.
[58] Ray L, May JC. Freeze drying/lyophilization of pharmaceutical and biological products. 3rd ed. New York: Informa Healthcare, 2010. ISBN 9781439825761. OCLC 664125915.
[59] https://project-pharmaceutics.com/services/lyophilization-process-development/?gclid=EAIaIQobChMIvePCss_R3wIV0_hRCh3gqg8OEAAYASAAEgJsUfD_BwE.

Fiona Mary Antony, Dharm Pal and Kailas Wasewar

2 Separation of bio-products by liquid–liquid extraction

Abstract: Solvent extraction one of the oldest approaches of separation known, remains one of the most well-known methods operating on an industrial scale. With the availability of variety of solvents as well as commercial equipment, liquid–liquid extractions finds applications in fields like chemicals and bio-products, food, polymer, pharmaceutical industry etc. Liquid–liquid extraction process is particularly suitable for biorefinery process (through conversion using microorganisms), featuring mild operational conditions and ease of control of process. The principles, types, equipment and applications of liquid–liquid extraction for bioproducts are discussed. Currently various intensification techniques are being applied in the field of liquid–liquid extraction for improving the process efficiency like hybrid processes, reactive extraction, use of ionic liquids etc, which are gaining importance due to the cost associated with the downstream processing of the fermentation products (20–50% of total production cost).

Keywords: bio-products, separation, reactive extraction, ionic liquid, biorefinery, equipments

2.1 Introduction

Bio-products or bio-based products commonly refer to chemicals, materials and energy resulting from renewable biological resources. The main category of bio-products is outlined in Figure 2.1. Bioenergy is produced, when biomass is treated by different physical, thermochemical biochemical, and other processes, in liquid (fuels like ethanol, biodiesel, bio-oil), solid (biomass) or gaseous (fuels like biogas and syngas) forms. Bio-composites, bio-fibres, bio-plastics, etc., can be included in the category of biomaterials. Industrial biochemical includes solvents, waxes, lubricants and adhesives manufactured from vegetable oils and organic chemicals (succinic acid, acetic acid, glycerol and methanol). These are important feed stocks for producing high-value, bio-based materials and biochemical, biopharmaceuticals such as vaccines, drugs and antibiotics, which have medicinal value and cosmetics. When comparing with petroleum-based counterparts, bio-products offers several advantages

This article has previously been published in the journal Physical Sciences Reviews. Please cite as: Wasewar, K., Antony, F. M., Pal, D. Separation of bio-products by liquid–liquid extraction *Physical Sciences Reviews* [Online] 2021, 2. DOI: 10.1515/psr-2018-0065

https://doi.org/10.1515/9783110574111-002

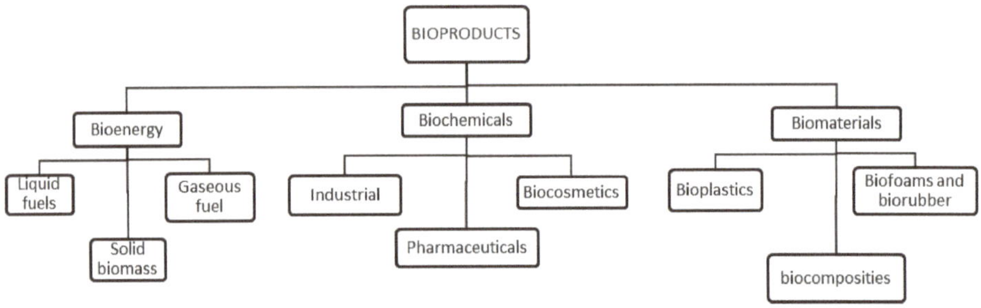

Figure 2.1: Typical classification of bio-products.

such as reduced greenhouse gas emissions, reduced toxicity and added biodegradability, sustainable production of renewable feedstock, etc.

There has been a growing interest in transformation of biomass to chemicals, energy as well as to biofuels. The factors that have led to it are reducing emission of greenhouse gases from fossil fuel, which contributes to global climate variation as well as considering factors like upsurge in global demand, price as well as reduction in potential obtainability of crude oil, and the requirement for energy security as well as independence. The global biomass-based chemical manufacture was assessed to be 50 million tons as of 2012 and further there was sale of $252 billion for biomass-based chemicals at 9% of the global chemical sales market [1]. The global market share of biomass centred chemicals is anticipated to rise from 2% in 2008 towards 22% in 2025 [2].

2.2 Separation and purification processes in biorefinery

A facility for converting biomass towards bio-products, which includes bioenergy (fuels, heat and power) as well as various range of co-products (comprising materials as well as chemicals) could be coined by the term bio-refinery. The conversion platform could be biochemical or thermo-chemical conversion approaches or a combination of both [3, 4]. The separation and purification of the desired products is one of the most vital constituents of bio-refineries as usually the cost associated with the separation as well as purification processes constitute 20–50% of the overall production costs of the bio-refineries. Most of the bio-production processes faces tremendous encounters in separation and purification due to numerous aspects involving product inhibition, low feed concentration, and low product yield. In the separation and purification in bio-refineries, there are numerous noteworthy challenges plus opportunities, to name a few-phytochemicals separation from biomass, biomass components separation (hemicellulose, lignin, cellulose and extractives, etc.), after preliminary pre-treatment and hydrolysis the separation as well as purification of various chemical species from the feed streams. Additionally, there are challenges involved in concentration of the desired species for different end product applications, along with simultaneous abstraction of products which are inhibitors in fermentation, from the spent pulping liquor, plus separation of lignin and chemicals. Finally, assimilation of separation and purification technologies with bioprocessing, along with separation as well as purification of downstream products [5]. For the effective commercialization of bio-refineries these opportunities and challenges needs to be considered. Separation and purification technologies in bio-refineries include equilibrium-based processes, for instance, absorption, distillation, supercritical fluid extraction, liquid–liquid extraction (LLE), affinity-based separation (adsorption, simulated moving bed and ion exchange), solid–liquid extraction, membrane separation and hybrid reaction–separation systems. The above-mentioned methods have its own merits and demerits.

2.3 Liquid–liquid extraction

LLE is a common mass-transfer operation for separating the constituents of liquid (the feed) by contacting with another liquid phase (the solvent) resulting in the production of a solvent-rich stream known as extract besides the residual liquid from which solute has been removed termed as raffinate. The LLE finds application in a number of processes including the retrieval of products from fermentation broth, pharmaceuticals food processing, industrial wastewater treatment, etc. This method takes benefit of variances in the chemical properties of the constituents of the feed to separate them, for instance, differences in polarity and hydrophobic/hydrophilic character. The constituents that transfer from one phase to other are driven by a deviance from the thermodynamic equilibrium, besides the interactions amongst the solvent phase and the feed components that determine the nature of the equilibrium state.

Extraction has been in practice since ancient time, way back to around 3500 BC for recovering products from several natural resources. Pharmaceutical oils, waxes and perfumes production have been documented at a Sumerian text dated 2100 BC. In medieval ages, extraction was used in hydrometallurgy field. The contemporary practice of LLE has its back ground in the middle to late nineteenth centuries while extraction turns into an significant laboratory technique, with the advances in thermodynamics and design of extraction apparatus. Industrial introduction of LLE technology occurred during a period from about 1920 to 1970. Nowadays LLE is a versatile technology with an extensive range of commercial uses mainly due to extensive variability of established process schemes as well as equipment options and the availability of a large number of commercial solvents and extractants.

Extraction offers several advantages over the direct separation techniques like distillation especially for heat-sensitive materials like penicillin and other antibiotics. For the dilute solutions, extraction is more economical, principally where water must be vaporized in distillation. Several pharmaceutical products like penicillin are formed in mixtures so complex that only liquid extraction is a viable separation method. Extraction technique offers flexibility in the choice of the operating conditions, by the proper choice of the amount and type of the solvent, separation effectiveness can be significantly improved.

An important step to effective separation by LLE is the selection of an appropriate solvent. Various solvent-selection criteria are outlined in Figure 2.2, of which some are desirable properties for separation (like solvent selectivity, density difference, recoverability), while the others will either advance the separation and/or make it furthercost-effective.

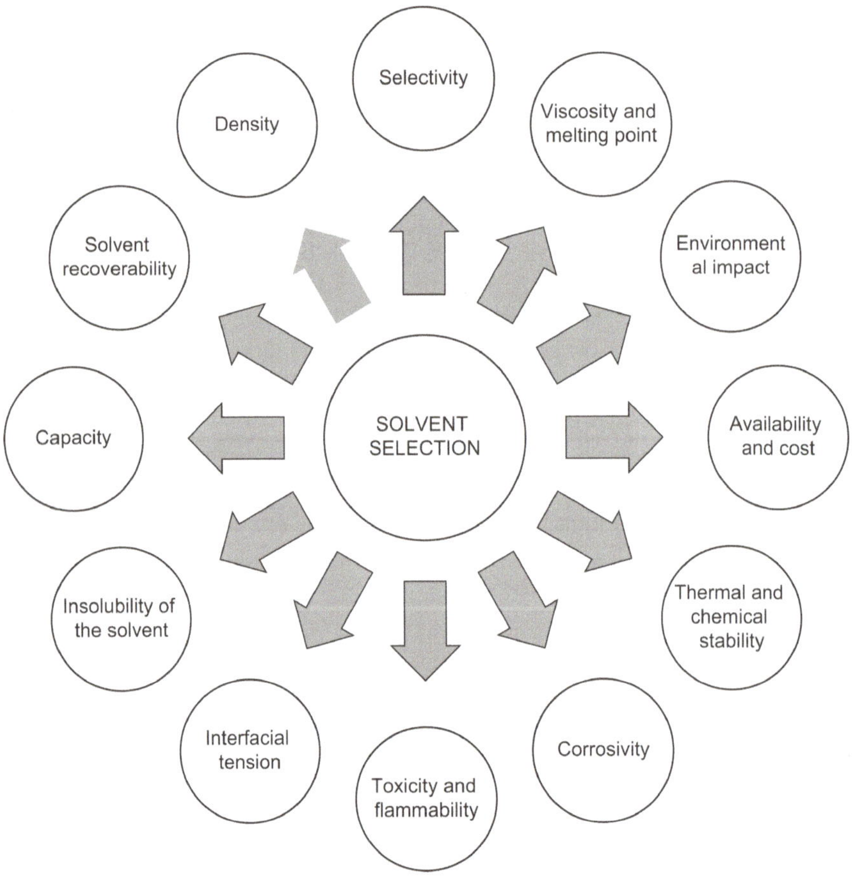

Figure 2.2: Solvent selection criteria.

The solvent should have higher choosiness towards the desired solute – a high separation factor value (which measures the comparative enhancement of solute one in the extract phase, related to other solute, after extraction) permits fewer stages to be used. A high value of distribution coefficient, K (the ratio of concentration of solute in the extract to that in the raffinate phase once equilibrium is accomplished), specifies a higher solvent capability for solute and allows lesser solvent to feed ratios. The solvent recovery should be comparatively easy (if possible by a simple flash or else a flash and subsequently a stripping column), which could be attained by use of a solvent with much lower or higher boiling temperature compared to the component chosen to be extracted. Further larger density difference can help in easy liquid phases settling. Solvents with higher viscosities lead to problems in pumping and dispersion and also reduce mass transfer efficiency. For pharmaceutical and food processing applications, nontoxic solvents should be preferred. Owing to an easier coalescence, higher interfacial tension permits a rapid settling, letting higher capacities. The solvent ought to be chemically as well as thermally stable should be non-corrosive as corrosive solvents increase equipment cost and requires treatment steps; it should be easily available, compatible with the environment as well as with upstream and downstream process steps.

2.4 Types of liquid–liquid extraction

LLE schemes practiced in industry can be categorized into different types, as discussed below.

2.4.1 Conventional extraction

Conventional extraction or standard extraction is also termed single-solvent extraction or simple extraction. It is the best commonly practiced extraction operation either using counter-current or cross-current flow of solvent, single-stage or multistage processing, and continuous or batch-wise operation. The process comprises transferring components from feed phase into a different phase. Except if the separation factor for the preferred solute with respect to undesirable solutes is much higher, this, method normally cannot achieve satisfactory solute purity.

2.4.2 Fractional extraction

This method combines solute recovery with co-solute rejection. For fractional extraction using dual-solvent, an extraction solvent as well as a wash solvent is utilized and comprises for product-solute recovery, at the raffinate end of the process a stripping section and for co-solute rejection and product purification, at the extract end of the process a washing section, thereby achieving high solute recovery and purity. It is normally used to retrieve aromatic constituents from crude hydrocarbon mixtures.

2.4.3 Dissociative extraction

This process system usually comprises partitioning of feeble bases or organic acids amongst an organic solvent and water. Here a sharp feed solutes separation is attained by considering the benefit of a variance in their pKa values. If the pKa dissimilarity is adequate, monitoring pH at a precise value can produce higher K values for one of the solute fraction and lower values for the other, thus letting a sharp separation.

2.4.4 pH-swing extraction

This method employs dissociative extraction ideas for recovering and purifying organic solutes that are ionisable, in a forward and back-extraction scheme, each extraction operation being done at an altered pH.

2.4.5 Reactive extraction

This process involves the development of a reversible extractant-solute interactions besides improved partition ratios for aiding anticipated separation. It involves combination of reaction as well as separation in the same unit operation, running a reaction in presence of two liquid phases besides taking benefit of the partitioning of reactants and products, between the two phases to increase reaction performance

2.4.6 Temperature-swing extraction

This process takes benefit of a variation in K value using temperature. K values can be mostly temperature sensitive when solvent-solute interactions in one or both phases include specific attractive interactions like hydrogen bonds or ion-pair bonds formation (such as in trialkyamine–carboxylic acid interactions) etc.

2.4.7 Membrane based solvent extraction

For the separation of solutes like metals, organic acids, etc., pertraction through liquid membranes (LM) besides membrane based solvent extraction (MBSE) have been employed [6–9]. MBSE an alternate to a conventional solvent extraction evades dispersing of the liquid phase which is connected with problems of emulsion formation as well as with the entrainment of the solvent droplets and its loss in many systems. LM offers good selectivity, easy handling besides low capital as well as operation costs also the extraction as well as stripping processes are joined in one step in LM. Bulk liquid membrane consists of liquid feed as well as stripping phase in contact with a LM phase.

2.4.8 Special extraction techniques

These techniques have been developed especially in the recovery of sensitive biological products, for enhancing the effectiveness of extraction.

2.4.8.1 Aqueous two-phase extraction

The aqueous two – phase extraction or aqueous bi-phasic extraction normally consist of the usage of a water-miscible polymer besides a salt (for instance Na_2SO_4 and PEG), or two in compatible water-miscible polymers [usually dextran and polyethylene glycol (PEG), a starch centred polymer], for forming two immiscible aqueous phases each containing 75+% water. For recovering of proteins as well as further biomolecules from broth or other aqueous feeds with negligible loss of activity, this technology provides mild conditions [10–12].

2.4.8.2 Reversed micellar extraction

For the isolation of proteins from an aqueous feed, this method uses microscopic water-in-oil micelles made by surfactants in addition to suspended in a hydrophobic organic solvent. The micelles are micro droplets of water with dimensions on the order of the protein designate to isolate, which provides a harmonious environment for the protein, permitting for its retrieval from a crude aqueous feed devoid of major protein activity loss [13, 14].

2.4.8.3 Supercritical fluid extraction

For extracting components from liquids or porous solids, this method consists of the use of light hydrocarbons or CO_2. This technique is often used for the purification of low-volume specialty chemicals or extraction of high value soluble components from natural materials. Applications include retrieval of active constituents from animal- and plant-derived feeds (e.g. vitamins from natural oils and flavour constituents), decaffeination of coffee etc. the advantages this process offers includes, mild operating temperatures (avoids product degradation), the solvent residues are nontoxic and can be easily removed, easy product recovery from extract fluid, solute separation factor could be attuned by making minor variations in the working pressure and temperature.

2.5 Applications of L-L extraction in bioprocess technology

LLE has several applications in bioprocess technology and some of them are outlined in Figure 2.3. A wide variety of chemicals as well as biofuels that are obtained through fermentation and algae are often retrieved and purified through LLE process. Phenolic compounds besides higher alcohols, considered as inhibitors during the fermentation of hydrolysed lignocellulosic biomass can be removed by this method. Carboxylic acids for instance lactic acid, tartaric acid, propanoic acid, citric acid, etc., obtained as fermentation product can be can be separated cost and energy efficiently by LLE. This method removes pollutants like phenols, aniline, nitrate aromatics from the wastewater. LLE finds applications in food industries like separation and purification of a particular flavour or fragrance. For the separation and purification of protein and peptides, LLE is a useful method, to their labile nature in the organic solvents. Factors like chemical properties of solvents, contact or mixing of the phase, extraction time etc affect the separation of protein.

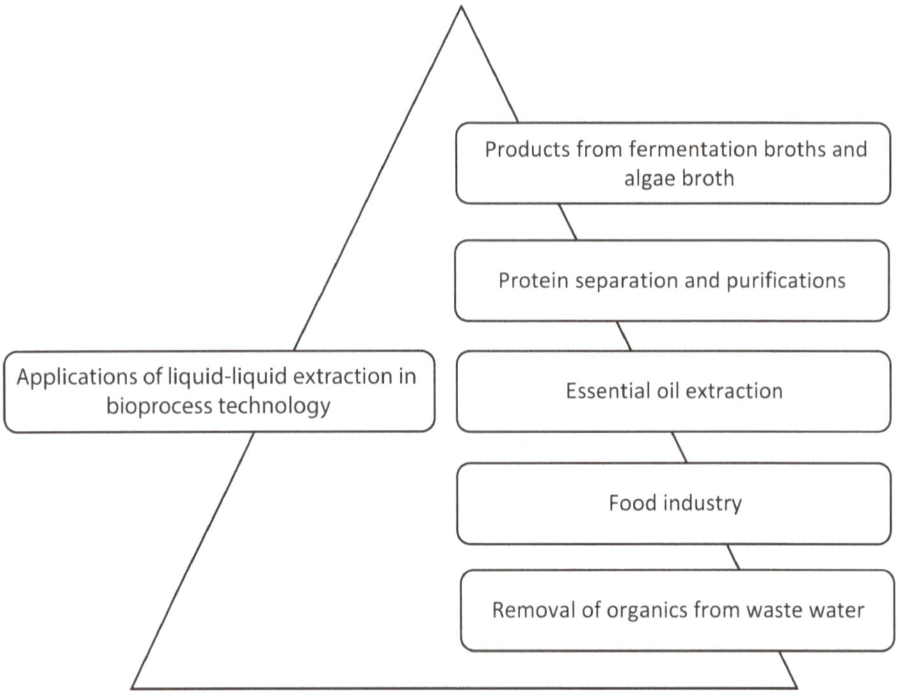

Figure 2.3: Few applications of LLE.

For separation of chemicals and biofuels from dilute liquid mixtures LLE could be used example includes, extraction of bio-alcohols [15] and carboxylic acids [16] from fermentation broths, inhibitor extraction (compounds which are toxic to fermentation microorganisms) from biomass hydrolysates [17], removal of impurities (soap, glycerol, methanol) in biodiesel from used cooking oils [18]. Reactive extraction can be used for organic acids extraction from fermentation broths for acids like maleic, succinic, itaconic, lactic acids etc.

One big challenge for new separation processes is retrieval of products from dilute streams in bio-refinery. Top potential platform chemicals from bio-refinery includes ethanol, glycerol and its derivatives, furans, succinic acid, levulinic acid, lactic acid, hydroxyl-propionic acid/aldehyde etc. and are discussed in [19, 20]. Table 2.1 gives information related to the separation of few of these chemicals by LLE.

Table 2.1: Few examples of application of extraction for separation of selected platform chemicals.

Chemical	Solvent used for extraction	reference
Ethanol	Aliphatic alcohol solvents	[21]
Ethanol	Isoamyl acetate, isooctyl alcohol, n-butyl acetate	[22]
Ethanol	Biobased oils, alcohols, esters	[23]
Ethanol	hexadecane, cottonseed oil, white light paraffin oil	[24]
Glycerol	Deep eutectic solvent choline-chloride: ethylene-glycol (molar ratio 1:2.5)	[25]
Glycerol	A Lewis basic mixture of quaternary ammonium salts with glycerol	[26]
succinic acid	1-octanol Primene JM-T1 (a C16–C22 primary amine,), n-Butyldiethanolamine, trioctylamine, (tris(2-ethylhexyl)amine	[27]
succinic acid	Piperidinium ionic liquids and phosphate salt	[28]
succinic acid	Amine–solvent mixtures	[29]
Levulinic acid-	Trioctyl phosphine oxide in, methyl isobutyl ketone	[30]
Levulinic acid-	Aliquat 336 in various organic solvents (benzene, dichloromethane, dodecane, methyl isobutyl ketone, 1-octanol)	[31]
Levulinic acid-	Tripropylamine in toluene	[32]
3-Hydroxypropionic acid	Tri-n-octylamine in n-decanol	[33]
3-Hydroxypropionic acid	Tri-*n* -octylamine and Aliquat 336 in n-decanol	[34]
lactic acid	Alamine 336 in methyl isobutyl ketone	[35]
lactic acid	Alamine 336 in decanol	[36]
lactic acid	Mixture of tripropylamine and trioctylamine in 1-octanol / n-heptane	[37]

2.6 Equipments for liquid–liquid extraction

In order to obtain higher mass transfer rates, the two phases in LLE must be brought to intimate contact, and further need to be separated. There are mainly two classes of equipment for solvent extraction, vessels with mechanical agitators for mixing, and mixing done by the flow of fluids. The equipments can be operated batch wise or continuously.

1. Mixer- settlers
2. Plate and agitated tower contactors
3. Packed and spray extraction towers

Several factors like high viscosity of feed, high solid content, low density difference between feed and solvent, etc., affect the biological extractions. Extractive fermentation process requires prior suspended solids separation by centrifugation or filtration and further the disposal of the filtered cake. Considerable amount of solute could be lost as part of filtrate during such separation. To overcome such problems, centrifugal contactors has been used widely in antibiotic extraction, in which counter current flow of phases facilitates mixing. Few classification of LLE equipments with application in chemical, pharmaceutical, fertilizer, food, metallurgical, petrochemical, nuclear field include static extraction columns, mixer-settlers, rotary-agitated columns, reciprocating-plate column, pulsed columns, centrifugal extractors, etc. [38–40].

2.7 New approaches

There are various approaches for advancements in the process of LLE process for bio-based and chemical processes. One such approach is instead of the conventional organic solvents, the use of new materials like ionic liquids, polymers, supra molecular structures, deep eutectic solvents, modifiers etc. that can escalate the extraction capability for bio-based products, for instance platform chemicals and biofuels, and fermentation-based products. Another approach is the development of novel extractors (like membrane extractors, micro-channels, micromixers) that when compared to traditional extractors, brings about a more efficient and effective extraction of products. The use of external force such as centrifugal and electric field, microwave, and ultrasound is also an effective step. The integration and hybrid separation methods, can also bring process intensification by means of pooled operations, reduced capital investment and energy consumption.

Hybrid extraction processes engage extraction operation in connotation with additional unit operation. For these operations, either the singular unit operations alone might not be capable of achieving all the separation goals, or the hybrid process may be more economical. Examples include extraction-distillation [41], extraction-crystallization [42], neutralization-extraction, reverse osmosis-extraction [43].

2.8 Reactive extraction

Normal LLE works at ambient or somewhat higher temperatures and re-extraction accomplished via a back wash or thermal strip. The solute transfer from one phase to another can be enhanced by means of "reacting" compounds dissolved in solvent phase (Figure 2.4). The usage of extractants diluted in solvents to enhance the selectivity has brought about different approaches in the arena of chemical industry (e.g. furfural extraction, inorganic and organic acids)and bio-chemical industry (e.g. amino acids, penicillin etc.), hydrometallurgy (e.g. metal mining) besides various environmental applications.

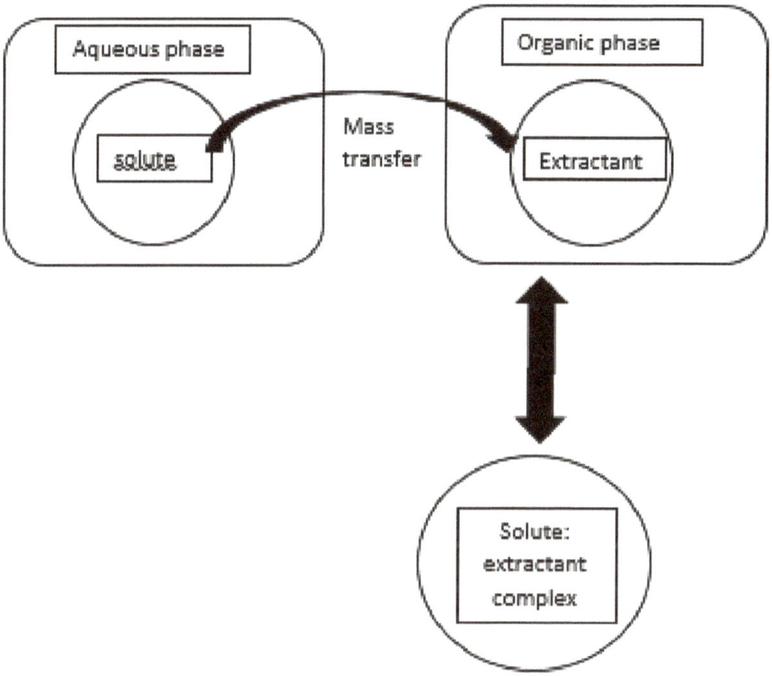

Figure 2.4: Reactive extraction of Solute from aqueous phase.

Reactive extraction is a process intensification technique, in which extraction is intensified through a mechanism involving a reversible reaction between the extractant and the extracted chemical species. Reactive extraction signifies an association between physical (diffusion as well as solubilization of system components) and chemical phenomena (extractant and solute reaction) [44]. "Reactive extraction" terms have been devised to classify extraction operations where either a chemical compound or an association complex is formed amongst the extractant and solute as an outcome of chemical or intermolecular interactions, respectively. These interactions could be symbolised by a reaction equation.

The extractants used for reactive extraction can be mainly divided into three extraction categories [45]: (i) carbon-bonded oxygen-bearing extractants (extraction by means of solvation); (ii) phosphorus-bonded, oxygen-bearing extractants (extraction via solvation) and (iii) aliphatic amines (extraction through ion pair formation or through proton transfer). Few important phosphorus as well as amine-based extractants are represented in Table 2.2. For practical reasons, the extractants are diluted in diluent, which are water immiscible, which offers the organic phase the desired physical properties (lower viscosity, higher interfacial tension, lower density), as most

Table 2.2: Some important phosphorus- and amine-based extractants and ionic liquids and their application for separation of few compound.

Extractants	Compound separated	Reference
Phosphorus– bonded oxygen bearing extractants		
Tributyl phosphate (TBP)	Itaconic, maleic, malic, oxalic, tartaric, and succinic	[46]
	Propionic	[47]
	Caproic	[48]
Tributyl phosphine oxide (TBPO)	Acetic and formic acids	[49]
Trioctyl phosphine oxide (TOPO)	Nicotinic acid	[50]
Di-(2-ethylhexyl)-phosphoric acid (D2EHPA)	Penicillin G	[51]
Tri-alkyl phosphine oxide (TRPO)	Glyoxylic, glycolic, acrylic, and benzoic acid	[52]
Aliphaticamine based extractants		
Lauryl-trialkylmethylamine (Amberlite LA-2)	Levulinic acid	[53]
	Glycolic acid	[54]
	Succinic acid	[55]
Tri-n-octylamine (Alamine 300)	Lactic acid	[56]
	Glutaric	[57]
	Citric	[58]
Tri-iso-octylamine (HOSTAREX A 324)	Citric, lactic, and malic	[59]
Tri-n-(octyl-decyl)-amine (Alamine 336)	Citric	[60]
	Acetic, lactic, succinic, malonic, fumaric, and maleic	[61]
Tri-n-dodeocylamine	Citric	[62]
	Formic	[63]
Ionic liquids		
Imidazolium ionic liquid: 1,3–dialkylimidazolium chloride	Alkaloids such as caffeine and nicotine	[64]
1-buty l,3-methylimidazolium tetrafluoroborate	Penicillin G	[65]
Phosphonium ionic liquid:- trihexyltetradecyl phosphonium di-2,4,4 trimethylpentylphosphinate	Butyric acid	[66]
Ammonium ionic liquid N,N-dimethylethanolamine propionate	Proteins	[67]
Pyrrolidinium ionic liquid:- 1-H-1- methyl pyrrolidinium chloride	Lignin	[68]

extractants are very viscous or even solid. Examples of few diluents used for extraction are given in Table 2.3.

Table 2.3: Examples of diluents used for reactive extraction.

Diluent category	Diluent example
Alcohols	2-Ethyl-1-hexanol, 1-octanol, 1-decanol
Ketone and ester	Methyl isobutyl ketone, diisobutyl ketone, butyl acetate
Alkyl aromatic	Toluene, xylene
Aliphatic hydrocarbon	Hexane, octane, dodecane
Natural oils	Sunflower oil, canola oil, rice bran oil

The use of conventional solvents and extractants in higher concentrations often create a negative impact on the activity of biochemical substances as well as problems like toxicity, volatility, flammability, difficult regeneration of the solvent, environmental pollution, etc. For making the separation process more environmentally friendly, substitution with ionic liquids, considered as green solvents is investigated owing to their superior properties over classical solvents. Ionic liquids are organic salts that remain liquid in an extensive temperature ranges with interesting and unique properties, which can be made from a huge range of anion and cation precursors to acquire optimum properties for particular applications. The most characteristic features of ionic liquids are negligible vapour pressure at room temperatures and thermal stability over a wide temperature range and exceptional solvent quality for numerous kinds of compounds, by a cautious selection of ions, their chemical/physical properties can be excellently adjusted, besides the properties specified above, the low toxic properties of ionic liquids make them further valued for today's industry. There are five elementary classes of ionic liquids according to cation structure- imidazolium, pyrrolidinium, phosphonium, ammonium, pyridinium and are mentioned in Table 2.2. The commonly used anions include hexafluorophosphates, halides, bis (trifluoromethylsulfonyl) imides, tetrafluoroborates, alkyl sulfates, methane sulfonates, etc. Ionic liquid finds various application in LLE of organic molecules, metal cations, and large biomolecules such as proteins.

Hydrophobic ionic liquids reveal to be potential extractants for carboxylic acids. There is an ongoing interest in ionic liquids application in extractive separations [69–74].

There are a numerous studies on organic acids reactive extraction in the literature investigating the numerous characteristics of reactive extraction like chemical interactions amongst extractants and acids, category of diluents and extractants, reaction mechanisms, consequence of aqueous and organic phase composition, effect of temperature and pH.

Reactive extraction finds applications in product recovery from fermentation broths including carboxylic acids for instance lactic, propanoic, citric, itaconic acid etc, alcohols like ethanol, or antibiotics like streptomycin or cephalosporin, penicillin. Fermentation broth composition affects the extraction of bioproducts. The impurities from the biocatalysts, left after acid solution separation affect phase separation behaviour and further reduce the extraction yield. While the process development for an extractive fermentation process, besides considering factors like temperature, extraction pH value, ionic strength, etc., the organic impurities also need to be considered, which requires additional process equipments like filtration units for solid separation.

2.8.1 Recovery of antibiotics

Antibiotics like penicillin, chloramphenicol, nisin, tylosin and erythromycin which are secreted by microbial cells can be retrieved from fermentation broth by extraction. Reactive extraction of penicillin G by means of quaternary amine with butyl acetate as solvent by centrifugal extraction in a hollow fibre liquid–liquid contactor resulted in faster extraction [75].

2.8.2 Recovery of carboxylic acids

Reactive extraction with particular extractants have been established to be a favourableas well as effective procedure for the separation of numerous carboxylic acids [76–79] like lactic acid [80–82] acrylic acid [83], benzoic acid [84], phenylacetic acid [85–88], tartaric acid [89], propionic acid [90–93], gallic acid [94–96], caproic acid [97–99], levulinic acid [100], nicotinic acid [101], protocatechuic acid [102–104], etc. Reactive extraction provides several advantages compared to the conventional approaches like the re-extraction of the acid, pH control of the reactor devoid of base addition, reusing of extractant, formation of reversible complex amongst the acid and extractant leading to increased separation efficiency, etc. Transfer of acid to the organic phase occurs chiefly by three mechanisms: acid ionisation in the aqueous phase, carboxylic acid partition amongst the organic phase and aqueous phase, dimerization in organic phase of the carboxylic acid.

There are several works on the extractive fermentation of carboxylic acids like lactic acid, butyric acid, etc. [105, 106]. Wu and Yang (2005) [107] produced butyric acid using immobilized *Clostridium tyrobutyricum* cells from glucose in a fibrous bed bioreactor and Alamine 336 in oleyl alcohol confined in a hollow fibre membrane extractor was used for its removal from the fermentation broth. The extractive fermentation process employing secondary amine extractant and hollow fibre contactor to separate selectively propionic acid from the fermentation broth was established to yield propionate from lactose by *Propionibacterium-acidi propionici* [108].

2.8.3 Other fermentation derived products

The retrieval of amino acids, commercially produced by microbial synthesis having basic characteristics like lysine, histidine, and arginine can be done using extraction. Extractants belonging to the category of high molecular weight quaternary aliphatic amines are generally used for the purpose of reactive extraction of amino acids. For the recovery of L-phenylanine amino acid, the use of integrated reactive extraction along with liquid–liquid centrifuge resulted in an increase in yield, lower by product formation etc [109].

LLE and reactive extraction finds application in separation of diols like 1,3-propanediol (PDO) from dilute aqueous fermentation broths, wherein PDO was converted to 2-methyl-1,2-dioxane and consequently extracted into o-xylene [110]. In another process for the recovery of PDO, butyr-aldehyde was used as the solvent, which gave improved product recovery. The most significant challenge in economic separation of PDO is its high cost of regeneration [111].

The recovery of bio-xylitol a substitute for sugar, obtained by enzymatic conversion of xylose has been reported by LLE with extraction yield of 78%, making it economically viable to carry out on an industrial scale [112, 113].

2.8.4 Extraction of cellular components and biopolymers

Biopolymers like proteins, peptides, cells and nucleic acids display different solubilities and hence partition differently in two phase aqueous systems, the common being PEG-K_2PO_4- water and PEG-dextran-water systems. Into the PEG phase, the protein partitions and contaminants like cell debris into the water rich phase, thus separation of protein can be achieved [114].

2.8.5 Biofuels

The most commonly used biofuel, ethanol was conventionally recovered by distillation, however the high temperature and energy needed for the process has led researchers to focus on liquid extraction for its separations. Reactive extraction using phosphonium based ionic liquids have been exploited for bioethanol separation from fermentation broths [115]. Another candidate to replace the fossil fuels, bio-butanol produced through ABE fermentation also has been successfully separated by reactive extraction by ionic liquids.

Biodiesel, which is produced by reaction of methanol and triglycerides, along with production of by-products glycerol and fatty acid methyl esters. Reactive extraction process combines the solvent extraction of oils and transesterification reaction leading to production of biodiesel.

2.8.6 Platform chemicals

Solvent extraction has been in practice for the separation of furfural an important chemical. Furan derivatives like 5-hydroxymethylfurfural, which has potential applications as building blocks for plastic production, has been separated using methyl isobutyl ketone in a biphasic system [116, 117]. Extraction of furfural from waste waters using ionic liquids has shown an efficiency up to 98% [118]. Another important platform chemical, for which reactive extraction has been employed is levulinic acid.

2.8.7 Biomass hydrolysate components and impurities

LLE is applied in bio-refinery domain, for the elimination of compounds from biomass feedstocks. In the initial step of biomass pre-treatment, a process known as detoxification process is done, wherein inhibitors present in biomass hydrolysates like formic acid, acetic acid, phenolic components etc. are extracted by means of organic solvents [119, 120].

2.8.8 Bio-products based on microalgae

In biodiesel using microalgal, literature reports the lipid extraction by solvent extraction, the use of supercritical fluids, pulsed electrical fields, and surfactant based *in situ* microalgal extraction [121].

2.9 Regeneration of solvent

For an effective process based on LLE, not only a good extraction but also an effective as well as easy retrieval of the solvent is required. Solvent regeneration is the "solvent treatment for re-cycling, e.g.by deduction of non-strippable solutes or degradation products" [122]. There are several methods in practice for the regeneration of solvent after extraction [123–125]. Some of the techniques for solvent regeneration includes regeneration by evaporation and stripping(direct evaporation can be employed in case of volatile acid extraction) [126, 127], by reactive back-extraction [128, 129], by physical back-extraction [130], temperature-swing regeneration [131], diluent-swing regeneration [132].

2.10 Conclusions

LLE is a favourable approach for the separation of various bio-based products like fermentation based products, biofuels, platform chemicals etc. The use of LLE for bio-product separation gained importance during the 20th century with the recovery of penicillin antibiotic from fermentation broth. Nowadays many process intensification techniques has come into practice like reactive separations, hybrid separations, use of alternative solvents, energy, etc., which has further led to an increase in the separation process by LLE.

References

[1] Biddy MJ, Scarlata C, Kinchin C. Chemicals from biomass: a market assessment of bioproducts with near-term potential. National Renewable Energy Laboratory (NREL). 2016. DOI:10.

[2] de Jong E. Bio-Based Chemicals: value added products from biorefineries. 2012. http://www.iea-bioenergy.task42-biorefineries.com/. May 2012.

[3] Huang HJ, Ramaswamy S, Tschirner UW, Ramarao BV. A review of separation technologies in current and future biorefineries. Sep Purif Techno. 2008;62:1–21.

[4] Huang, H.-J. and Ramaswamy S. Separation and purification of phytochemicals as co-products in biorefineries. In: Biorefinery Co-Products: Phytochemicals, Primary Metabolites and Value-Added Biomass Processing, edited by Bergeron C., D. Julie Carrier, and Ramaswamy S. John Wiley & Sons, Ltd, U S A, 2012.

[5] Ramaswamy S, Huang H. J, & Ramarao, B. V. Separation and purification technologies in biorefineries. John Wiley and Sons, U S A, 2013. https://doi.org/10.1002/9781118493441.

[6] Schlosser Š, Sabolová E, Kertész R, Kubišová L. Factors influencing transport through liquid membranes and membrane based solvent extraction. J Sep Sci. 2001;24:509–18.

[7] Giorno L, Spicka P, Drioli E. Downstream processing of lactic acid by membrane-based solvent extraction. Sep Sci Technol. 1996;31:2159–69.

[8] Prasad R, Sirkar K.K. (1992) Membrane-based solvent extraction. In: Ho W.S.W., Sirkar K.K. (eds) Membrane handbook. Springer, Boston, MA.1992:727-63. https://doi.org/10.1007/978-1-4615-3548-5_41.

[9] Schlosser Š. Membrane Based Processes with Immobilized Interface. In: ibid, Kluwer Academic, 2000:55–72.

[10] Walter H, Johansson G. editor. Aqueous two phase systems, methods in enzymology. vol. 228. Elsevier, Academic Press,U S A, 1994.

[11] Zaslavsky: BY. Aqueous two-phase partitioning – physical chemistry and bioanalytical applications. New York, Basel, Oxford: Marcel Dekker, Inc., 1995. ISBN 0-8247-9461-3.

[12] Clark B. Chap. 6 in Biochemical Engineering Dekker. 1997. 474–82.

[13] Ayala CA, Kamat S, Beckman AJ. Protein extraction and activity in reverse micelles of a nonionic detergent. Biotechnol Bioeng. 1992;39:806–14.

[14] Bordier C. Phase separation of integral membrane proteins in Triton X-114 solution. J Biolog Chem. 1981;256:1604–7.

[15] Simoni LD, Chapeaux A, Brennecke JF, Stadtherr MA. Extraction of biofuels and biofeedstocks from aqueous solutions using ionic liquids. Comput Chem Eng. 2010;34:1406–12.

[16] Bressler E, Braun S. Separation mechanisms of citric and itaconic acids by water-immiscible amines. J Chem Technol Biotechnol. 1999;74:891–6.

[17] Grzenia D, Qian X, da Silva SS, Wang X, Wickramasinghe SR. Membrane extraction for biofuel production. In: Membrane science and technology. vol. 14. Edited by Oyama S.T, Susan M. Williams S, Elsevier B.V, 2011:213–33. DOI:10.1016/B978-0-444-53728-7.00010-0.

[18] Berrios M, Mart´ın MA, Chica AF, Mart´ın A. Purification of biodiesel from used cooking oils. Appl Energy. 2011;88:3625–31.

[19] Werpy T, Petersen GE. TopValueAddedChemicalsfromBiomass, volume I. ResultsofScreeningforPotentialCandidatesfromSugarsandSynthesisGas. 2004:76.
[20] Bozell JJ, Petersen GR. Technology development fortheproductionofbiobasedproductsfrombiorefinerycarbohydrates-the US department ofEnergy's "Top 10" revisited. Green Chem. 2010;12:539–54.
[21] Offeman RD, Stephenson SK, Franqui D, Cline JL, Robertson GH, Orts WJ. Extraction of ethanol with higher alcohol solvents and their toxicity to yeast. Sep Purif Technol. 2008;63:444–51.
[22] Koullas DP, Umealu OS, Koukios EG. Solvent selection for the extraction of ethanol from aqueous solutions. Sep Sci Technol. 1999;34:2153–63.
[23] Offeman RD, Stephenson SK, Robertson GH, Orts WJ. Solvent extraction of ethanol from aqueous solutions using biobased oils, alcohols, and esters. J Am Oil Chem Soc. 2006;83:153–7.
[24] Mehta GD, Fraser MD. A novel extraction process for separating ethanol and water. Ibid. 1985;24:556–60.
[25] Petracic A, Sander A, Magic L. Separation of free glycerol and glycerides from biodiesel by means of liquid-liquid extraction. Sci J Energy Eng. 2017;5:87–94.
[26] Abbott AP, Cullis PM, Gibson MJ, Harris RC, Raven E. Extraction of glycerol from biodiesel into a eutectic based ionic liquid. Green Chem. 2007;9:868.
[27] Moraes LDS, Kronemberger FDA, Ferraz HC, Habert AC. Liquid–liquid extraction of succinic acid using a hollow fiber membrane contactor. J Ind Eng Chem. 2015;21:206–11.
[28] Lee WY, Hong YK. Liquid extraction of succinic acid by aqueous two-phase systems composed of piperidinium ionic liquids and phosphate salt. Korean J Chem Eng. 2016;54:52–6.
[29] Kurzrock T, Weuster-Botz D. New reactive extraction systems for separation of bio-succinic acid. Bioprocess Biosyst Eng. 2011;34:779–87.
[30] Brouwer T, Blahusiak M, Babic K, Schuur B. Reactive extraction and recovery of levulinic acid, formic acid and furfural from aqueous solutions containing sulphuric acid. Sep Purif Technol. 2017;185:186–95.
[31] Datta D, Marti ME, Pal D, Kumar S. Equilibrium study on the extraction of levulinic acid from aqueous solution with aliquat 336 dissolved in different diluents: solvent's polarity effect and column design. Chem Eng Data. 2017;62:3–10.
[32] Uslu H. Reactive extraction of levulinic acid using TPA in toluene solution: LSER modeling, kinetic and equilibrium studies. Sep Sci Technol. 2008;43:1535–48.
[33] Chemarin F, Moussa M, Chadni M, Pollet B, Lieben P, Allais F, et al. New insights in reactive extraction mechanisms of organic acids: an experimental approach for 3-hydroxypropionic acid extraction with tri-n-octylamine. Sep Purif Technol. 2017;179:523–32.
[34] Moussa M, Burgé G, Chemarin F, Bounader R, Saulou-Bérion C, Allais F, et al. Reactive extraction of 3-hydroxypropionic acid from model aqueous solutions and real bioconversion media. comparison with its isomer 2-hydroxypropionic (lactic) acid. J Chem Technol Biotechnol. 2015;91:2276–85.
[35] Wasewar KL, Heesink AB, Versteeg GF, Pangarkar VG. Reactive extraction of lactic acid using alamine 336 in mibk : equilibria and kinetics. J Biotech. 2002;97:59–68.
[36] Wasewar KL, Heesink AB, Versteeg GF, Pangarkar VG. Equilibria and kinetics for reactive extraction of lactic acid using alamine 336 in decanol. J Chem Tech Biotech. 2002a;77:1068–75.
[37] Hong YK, Hong WH. Reactive extraction of lactic acid with mixed tertiary amine extracts. Biotechnol Tech. 1999;13:915–8.
[38] Perry RH, Green DW. Perry's chemical engineer's handbook. 7th ed. New York: McGraw-Hill; 1997.

[39] McCabe WL, Smith JC, Harriott P. Unit operations of chemical engineering. 5th ed. McGraw Hill N.Y., 1993.
[40] Seader JD, Henley EJ. Separation process principles. John Wiley, U S A 1998.
[41] Nhien LC, Long NV, Lee M. Process design of hybrid extraction and distillation processes through a systematic solvent selection for furfural production. Energy Procedia. 2017;105:1084–9.
[42] Maeda K, Nomura Y, Fukui K, Hirota S. Separation of fatty acids by crystallization using two liquid phases. Korean J Chem Eng. 1997;14:175–8.
[43] Thompson JA. Selective extraction solvent recovery using regenerated cellulose membrane under reverse osmosis conditions. 1985:US4510047A.
[44] Kumar S, Wasewar KL, Babu BV. Intensification of nicotinic acid separation using organophosphorus solvating extractants: extraction equilibria. Chem EngTechnol. 2008;31:1584–90.
[45] Kertes AS, King CJ. Extraction chemistry of fermentation product carboxylic acids. Bio Technol Bioeng. 1986;28:269–82.
[46] Kyuchoukov G, MoralesA F, Albet J, Malmary G, Molinier J. On the possibility of predicting the extraction of dicarboxylic acids with tributylphosphate dissolved in a diluent. J Chem Eng Data. 2008;53:639–47.
[47] Kumar S, Datta D, Babu BV. Differential evolution approach for reactive extraction of propionic acid using tri-*n*-butyl phosphate (TBP) in kerosene and 1-decanol. Mater Manuf Process. 2011;26:1222–8.
[48] Wasewar KL, Shende D. Equilibrium study for reactive extraction of caproic acid in MIBK and Xylene. Engineering. 2011;3:829–35.
[49] Wardell JM, King CJ. Solvent equilibriums for extraction of carboxylic acids from water. J Chem Eng Data. 1978;23:144–8.
[50] Kumar S, Wasewar KL, BabuB V. Intensification of nicotinic acid separation using organophosphorous solvating extractants by reactive extraction. ChemEng Technol. 2008;31:1584–90.
[51] Ren Z, Zhang W, Li J, Wang S, Liu J, Lv Y. Effect of organic solutions on the stability and extraction equilibrium of penicillin G, J. Chem Eng Data. 2010;55:2687–94.
[52] Li Y, Wang Y, Li Y, Dai Y. Extraction of glyoxylic acid, glycolic acid, acrylic acid, and benzoic acid with trialkylphosphine oxide, J. Chem Eng Data. 2003;48:621–4.
[53] Uslu H, Kirbaşlar Şl, Wasewar KL. Reactive extraction of levulinic acid by amberlite LA-2 extractant. J Chem Eng Data. 2009;54:712–8.
[54] Asci YS, Inci I. Extraction of glycolic acid from aqueous solutions by amberlite la-2 in different diluent solvents. J Chem Eng Data. 2009;54:2791–4.
[55] Asci YS, İnci İ, Uslu H. LSER modelim extraction of succinic acid by tridodecyl amine dissolved in 2-octanone and 1-octanlol. J Ind Eng Chem. 2010;32:1951–7.
[56] Juang RS, HuangR H. Kinetic studies on lactic acid extraction with amine using a microporous membrane-based stirred cell. J Membr Sci. 1997;129:185–96.
[57] Pehlivanoğlu N, Uslu H, KirbaşlarŞ I. Experimental and modeling studies on the extraction of glutaric acid by trioctylamine. J Chem Eng Data. 2009;54:3202–7.
[58] Keshav A, Norge P, WasewarK L. Reactive extraction of citric acid using tri-n-octylamine in nontoxic natural diluents: part 1– equilibrium studies from aqueous solutions. Appl Biochem Biotechnol. 2012;167:197–213.
[59] Malmary G, Albet J, Putranto A, Hanine H, Molinier J. Recovery of aconitic and lactic acids from simulated aqueous effluents of the sugar-cane industry through liquid-liquid extraction. J Chem Technol Biotechnol. 2000;75:1169–73.

[60] Bízek V, Horáček J, Řeřicha R, Koušová M. Amine extraction of hydroxycarboxylic acids. 1. extraction of citric acid with 1-octanol/n-heptane solutions of trialkylamine. Ind Eng Chem Res. 1992;31:1554–62.
[61] Tamada JA, Kertes AS, KingC J. Extraction of carboxylic acids with amine extractants. 1. equilibria and law of mass action modeling. Ind Eng Chem Res. 1990;29:1319–26.
[62] Bayazit ŞS, Uslu H, Inci I. Comparative equilibrium studies for citric acid by amberlite LA-2 or Tridodecylamine (TDA). J Chem Eng Data. 2009;54:1991–6.
[63] Şahin S, Bayazit ŞS, Bilgin M. Incil.Investigation of formic acid separation from aqueous solution by reactive extraction: effects of extractant and diluent. J Chem Eng Data. 2010;55:1519–22.
[64] Freire MG, Neves CM, Marrucho IM, Lopes JN, Rebelob LP, Coutinho JA. High-performance extraction of alkaloids using aqueous two- phase systems with ionic liquids. Green Chem. 2010;12:1715–18.
[65] Lui Q, Xuesheng H, Wang Y, Yang P, Xia H, Yu J, et al. Extraction of penicillin G by aqueous two-phase system of [Bmim]BF4/NaH2PO4. Chin Sci Bull. 2005;50:1582–5.
[66] Marták J, Schlosser Š. New mechanism and model of butyric acid extraction byphosphonium ionic liquid. J Chem Eng Data. 2016;61:2979–96.
[67] Chen J, Wang Y, Zeng Q, Ding X, Huang Y. Partition of proteins with extraction in aqueous two-phase system by hydroxyl ammonium-based ionic liquid. Anal Methods. 2014;6:4067–76.
[68] Hu XM, Li S, Ma HH, Zhang BX, Gao YF. Pyrrolidinium ionic liquids as effective solvents for lignin extraction and enzymatic hydrolysis of lignocelluloses. Bio Res. 2016;11:7672–85.
[69] Liu JF, Jiang GB, J¨onsson J˚A. Application of ionic liquids in analytical chemistry. TrAC-Trends Anal Chem. 2005;24:20–7. DOI:10.1016/j.trac.2004.09.005.
[70] Marták J, Schlosser Š. Phosphonium ionic liquids as new, reactive extractants of lactic acid. Chem Pap. 2006;60:395–8.
[71] Marták J, Schlosser Š. Extraction of lactic acid by phosphonium ionic liquids. Sep Purif Technol. 2007;57:483–94.
[72] McFarlane WB, Ridenour H, Luo RD, Hunt DD, Ren RX. Room temperature ionic liquids for separating organics from produced water. Sep Sci Technol. 2005;40:1245–65.
[73] Pandey S. Analytical applications of room-temperature ionic liquids: A review of recent efforts. Analytica Chimica Acta. 2006;556:38–45.
[74] Zhao H, Xia SQ, Ma PS. Use of ionic liquids as 'green' solvents for extractions. J Chem Technol Biotechnol. 2005;80:1089–96.
[75] Schügerl K. Solvent extraction in biotechnology recovery of primary and secondary metabolites. Berlin Heidelberg: Springer-Verlag; 1994.
[76] Morales M, Ataman M, Badr S, Linster S, Kourlimpinis I, Papadokonstantakis S, et al. Sustainability assessment of succinic acid production technologies from biomass using metabolic engineering. Energy Environ Sci. 2016;9:2794–805.
[77] Kaur G, Elst K. Development of reactive extraction systems for itaconic acid: a step towards in situ product recovery for itaconic acid fermentation. RSC Adv. 2014;4:45029–39.
[78] Wardell JM, King CJ. Solvent equilibria for extraction of carboxylic acids from water. J Chem Eng Data. 1978;23:144–8.
[79] Wennersten R. The extraction of citric acid from fermentation broth using a solution of a tertiary amine. J Chem Technol Biotechnol. 1983;33:85–94.
[80] Wasewar KL, Heesink AB, Versteeg GF, Pangarkar VG. Equilibria and kinetics for back extraction of lactic acid using trimethylamine. J Chem Eng Sci. 2004;59:2315–20.
[81] Wasewar KL, Yawalkar A, Moulijn J, Pangarkar VG. Fermentation of glucose to lactic acid coupled with reactive extraction: A review. Ind Eng Chem Res. 2004;43:5969–82.

[82] Wasewar KL, Heesink AB, Versteeg GF, Pangarkar VG. Intensification of enzymatic conversion of glucose to lactic acid by reactive extraction. Chem Eng Sci. 2003;58:3385–94.
[83] Keshav A, Wasewar KL. Reactive extraction of acrylic acid using Tri-n-Butyl phosphate in different solvents. J Chem Eng Data. 2009;54:1782–6.
[84] Datta D, Kumar S, Wasewar KL. Reactive extraction of benzoic acid and pyridine –3-carboxylic acid using organophosphoric and aminic extractant dissolved in binary diluent mixtures. J Chem Eng Data. 2011;56:3367–75.
[85] Athankar KK, Wasewar KL, Varma MN, Shende DZ. Separation of phenylacetic acid using tri n-butyl phosphate in hexanol: Equilibrium and kinetics. Sep Sci Technol. 2017;52:2696–703.
[86] Athankar KK, Varma MN, Shende DZ, Yoo CK, Wasewar KL. Reactive extraction of phenylacetic acid with tri-n-butyl phosphate in benzene, hexanol, and rice bran oil at 298 K. J Chem Eng Data. 2013;58:3240–8.
[87] Athankar KK, Wasewar KL, Varma MN, Shende DZ, Uslu H. Extractive separation of benzylformic acid with phosphoric acid tributyl ester in ccl4, decanol, kerosene, toluene, and xylene at 298 K. J Chem Eng Data. 2015;60:1014–22.
[88] Gaidhani HK, Wasewar KL, Pangarkar VG. Intensification of enzymatic hydrolysis of penicillin-g: 1. equilibria and kinetics of extraction of phenyl acetic acid by alamine 336. Chem Eng Sci. 2002;57:1979-.
[89] Sharma H, Singh K, Wasewar KL, Athankar KL. L(+)-tartaric acid separations using aliquat 336 in n-heptane, kerosene, and 1-octanol at 300 ± 1 K. J Chem Eng Data. 2017;62:4047–63.
[90] Keshav A, Wasewar KL. Back extraction of propionic acid from loaded organic phase. Chem Eng Sci. 2010;65:2751–7.
[91] Keshav A, Wasewar KL, Chand S. Reactive extraction of propionic acid using tri-n-butyl phosphate in petroleum ether: equilibrium study. Chem Biochem Eng Q. 2008;22:433–7.
[92] Keshav A, Wasewar KL, Chand S. Reactive extraction of propionic acid using tri-n-octylamine. Chem Eng Commun. 2009;197:606–26.
[93] Keshav A, Wasewar KL, Chand S. Equilibrium studies for extraction of propionic acid using tri-n-butyl phosphate in different solvents. J Chem Eng Data. 2008;22:1424–30.
[94] Athankar KK, Wasewar KL, Varma MN, Shende DZ. Reactive extraction of gallic acid with tri-n-caprylylamine. New J Chem. 2016;40:2413–7.
[95] Rewatkar K, Shende DZ, Wasewar KL. Effect of temperature on reactive extraction of gallic acid using tri-n-butyl phosphate, tri-n-octylamine and aliquat 336. J Chem Eng Data. 2016;61:3217–24.
[96] Rewatkar K, Shende DZ, Wasewar KL. Reactive separation of gallic acid: experimentation and optimization using response surface methodology and artificial neural network. Chem Biochem Eng Q. 2017;31:33–46.
[97] Wasewar KL, Shende DZ. Equilibrium for the reactive extraction of caproic acid using tri-n-butyl phosphate in methylisobutyl ketone and xylene. J Chem Eng Data. 2011;56:3318–22.
[98] Wasewar KL, Shende DZ. Reactive extraction of caproic acid using tri-nbutyl phosphate in hexanol, octanol, and decanol. J Chem Eng Data. 2011;56:288–97.
[99] Wasewar KL, Shende DZ. Extraction of caproic acid using tri-n-butyl phosphate in benzene and toluene at 301 K. J Chem Eng Data. 2010;55:4121–5.
[100] Uslu H, Kirbaslar SI, Wasewar KL. Reactive extraction of levulinic acid by amberlite la-2 extractant. J Chem Eng Data. 2009;54:712–8.
[101] Waghmare MD, Wasewar KL, Sonawane SS, Shende DZ. Reactive extraction of picolinic acid and nicotinic acid by naturalnon-toxic solvent. Sep Purif Technol. 2013;120:296–303.
[102] Antony FM, Wasewar KL, BS D. Efficacy of tri-n-octylamine, tri-n-butyl phosphate and di-(2-ethylhexyl) phosphoric acid for reactive separation of protocatechuic acid. Sep Sci Technol. 2018c. DOI:https://doi.org/10.1080/01496395.2018.1556692.

[103] Antony FM, Wasewar KL. Separation of protocatechuic acid using di-(2-ethylhexyl)phosphoric acid in isobutyl acetate, toluene, and petroleum ether. J Chem Eng Data. 2018a;207. DOI:10.1021/acs.jced.7b00797.

[104] Antony FM, Wasewar KL. Reactive separation of protocatechuic acid using Tri-n-octyl amine and Di-(2-ethylhexyl) phosphoric acid in methyl isobutyl ketone. Sep Purif Technol. 2018b. DOI:10.1016/j.seppur.2018.06.037.

[105] Yang S-T, Huang H, Tay A, Qin W, De Guzman L, San Nicolas EC. Extractive fermentation for the production of carboxylic acids. In: Yang S-T, editor. Bioprocessing for value added products from renewable resources. Elsevier, U S A, 2007:421–46.

[106] López-Garzón CS, Straathof AJJ. Recovery of carboxylic acids produced by fermentation. Bio Technol Adv. 2014;32:873–904.

[107] Wu ZT, Yang ST. Extractive fermentation for butyricacid production from glucose by Clostridiumtyrobutyricum. Biotechnol Bioeng. 2003;82:93–102.

[108] Jin ZW, Yang ST. Extractive fermentation for enhanced propionic-acid production from lactose by propionibacteriumacidipropionici. Biotechnol Prog. 1998;14:457.

[109] Rüffer N, Heidersdorf U, Kretzers I, Sprenger GA, Raeven L, Takors R. Fully integrated L-phenylalanine separation and concentration using reactive-extraction with liquid-liquid centrifuges in a fed-batch process with E. coli. Bioprocess Biosyst Eng. 2004;26:239–48.

[110] Xiu Z-L, Zeng A-P. Present state and perspective of downstream processing of biologically produced 1,3-propanediol and 2,3-butanediol. Appl Microbiol Biotechnol. 2008;78:917–26. DOI:10.1007/s00253-008-1387-4.

[111] Hao J, Xu F, Liu H, Liu D. Downstream processing of 1,3-propanediol fermentation broth. J Chem Technol Biotechnol. 2006;81:102–8.

[112] Mussatto SI, Santos JC, Filho WC. Silva SS purification of xylitol from fermented hemicellulosichydrolyzate using liquid-liquid extraction and precipitation techniques. Bio Technol Lett. 2005;15:1113–15.

[113] Mun W, Islam L, Rafiqul SM, Munaim A, Sakinah M, Zularisam W. Purification of bioxylitol by liquid-liquid extraction from enzymatic reaction mixture. Sep Sci Technol. 2016. DOI:10.1080/01496395.2016.1203335.

[114] Seader JD, Henley EJ, Roper DK. Separation process principles: chemical and biochemical operations. USA: John Wiley & Sons, Inc., 2006.

[115] Freire MG, Neves CM, Marrucho IM, Lopes JN, Rebelo JP, Coutinhob JA. High-performance extraction of alkaloids using aqueous two-phase systems with ionic liquids. Green Chem. 2011;13:1517.

[116] Rosatella AA, Simeonov SP, Fradea RF, Afonso CA. 5-Hydroxymethylfurfural (HMF) as a building block platform: Biological properties, synthesis and synthetic applications. Green Chem. 2011;13:754.

[117] Teong SP, Yia G, Zhang Y. Hydroxymethylfurfural production from bioresources: past, present and future. Green Chem. 2014;16:2015.

[118] Habbal S, Haddou B, Kameche M, Derriche Z, Canselier& JP, Gourdon C. Cloud point or ionic liquid extraction of furfural from aqueous solution: a comparative study based upon experimental design. Desalin Water TreaT. 2016;57:23770–8. DOI:10.1080/19443994.2015.1133322.

[119] Kiss A, Costin B. Intensified Downstream Processing in Biofuels Production. 2018. 10.1039/9781788010320-00062.

[120] Jönsson LJ, Martín C. Pretreatment of lignocellulose: Formation of inhibitory by-products and strategies for minimizing their effects. Bioresour Technol. 2016;199:103–12. DOI:10.1016/j.biortech.2015.10.009.

[121] Jessop PG, Heldebrant DJ, Li X, Eckert CA, Liotta CL. Reversible nonpolar-to-polar solvent. Nature. 2005;436:1102.

[122] IUPAC. Compendium of Chemical Terminology. 2nd ed. (the "Gold Book"). Compiled by A. D. McNaught and A. Wilkinson. Blackwell Scientific Publications, Oxford (1997). Online version (2019-) created by S. J. Chalk. ISBN 0-9678550-9-8. https://doi.org/10.1351/goldbook.

[123] Blahušiak M, Schlosser Š, Cvengroš J. Simulationof a new regeneration process of solvents with ionic liquid by short-path distillation. Sep Purif Technol. 2012;97:186–94.

[124] Reyhanitash E, Zaalberg B, Ijmker HM, Kersten SR, Schuur B. CO2-enhanced extraction of acetic acid from fermented waste water. Green Chem. 2015;17:4393–400.

[125] Reyhanitash E, Fufachev E, vanMunster KD, vanBeek MB, Sprakel LM, Edelijn CN, et al. Recovery and conversion of acetic acid from a phosphonium phosphinate ionic liquid to enable valorization of fermented waste water. Green Chem. 2019;21:2023–34.

[126] Blahušiak M, Schlosser Š, Cvengroš J, Marták J. New approach to regeneration of an ionic liquid containing solvent by molecular distillation. Chem Pap. 2011;65:603–7.

[127] Reyhanitash E, Fufachev E, Van Munster KD, Van Beek MB, Sprakel LM, Edelijn CN, et al. Recovery and conversion of acetic acid from a phosphonium phosphinate ionic liquid to enable valorization of fermented wastewater. Green Chem. 2019. DOI:10.1039/c9gc00725c.

[128] Siebold MP. Comparison of the production of lactic acid by three different lactobacilli and its recovery by extraction and electrodialysis. Process Biochem. 1995; 30: 81–95.

[129] Malmary G, Albet J, Putranto A, Hanine H, Molinier J. Measurement of partition coefficients of carboxylic acids between water and triisooctylamine dissolved in various diluents. J Chem Eng Data. 1998;43:849–51.

[130] Kroon MC, van Spronsen J, Peters CJ, Sheldon RA, Witkamp G-J. Recovery of pure products from ionic liquids using supercriticalcarbon dioxide as a co-solvent in extractions or as an anti-solvent in precipitations. Green Chem. 2006;8:246–9.

[131] Brouwer T, Blahusiak M, Babic K, Schuur B. Reactive extraction and recovery of levulinic acid, formic acidand furfural from aqueous solutions containing sulphuric acid. Sep Purif Technol. 2017;185:186–95.

[132] Krzyżaniak A, Leeman M, Vossebeld F, Visser TJ, Schuur B, de Haan A. Novel extractants for the recovery of fermentation derivedlactic acid. Sep Purif Technol. 2013;111:82–9.

José Coelho, Paolo Trucillo, Beatriz Nobre, António Figueiredo Palavra, Roberta Campardelli and Ernesto Reverchon

3 Extraction and bioprocessing with supercritical fluids

Abstract: Supercritical fluid (SCF) technologies have emerged as a real alternative to various natural product extraction processes and pharmaceutical production to obtain micronized particles, coprecipitates, nanocomposite polymer structures and liposomes, in addition to other increasingly larger applications described in literature. In the present work, a brief literature review of the application of supercritical fluid extraction (SFE) is presented. This is evidenced by several publications and patents, contributions from several countries and the increase of industries around the world dedicated to this technique. Next, we aim to focus the analysis of SFE on a review of the literature applied to microalgae as a substitute primitive feedstock due to its high growth rate, valuable biologically active lipophilic substances, and photosynthetic efficiency without competition with food sources or needs of arable lands. We finally discussing an SCF bioprocess with a very new perspective for liposome production focalized on its potential at industrial scale.

3.1 Introduction

The use of supercritical fluids (SCFs) in several processes has advantages related to the transport properties, such as high diffusivity and low viscosity, which can potentiate mass transfer phenomena. In addition, the variation of pressure and/or temperature allowed to adjust solvent properties, providing high flexibility in the extraction, reaction or encapsulation processes. Among the SCFs used at laboratorial and industrial scale, carbon dioxide (SC-CO_2) is the most common because of its advantages such as low cost, non-toxicity, non-flammability, inertness, possibility of total recovery and moderate critical properties ($Pc = 7.38\,MPa$, $Tc = 304.2\,K$), when compared to other potential green solvents [1–3]. A SCF shows an extensive range of solvation power as its density is strongly dependent to upon temperature and pressure. Small temperature change or pressure can change a compound's solubility in a SCF by an order of magnitude or more. Moreover, selectivity of nonpolar SCF (like is the case of SC-CO_2) can also be improved by addition of modifiers (entrainers or co-solvents), which are typically polar organic solvents e. g. acetone,

This article has previously been published in the journal Physical Sciences Reviews. Please cite as: Coelho, J., Trucillo, P., Nobre, B., Figueredo Palavra, A., Campardelli, R., Reverchon, E. Extraction and bioprocessing with supercritical fluids *Physical Sciences Reviews* [Online] 2020, 9. DOI: 10.1515/psr-2018-0069

https://doi.org/10.1515/9783110574111-003

ethanol, methanol. Changing the amount of modifier allows extensive latitude in the distinction of solvent power.

The use of SCFs has expanded seriously since one of the first article was published in 1969 [4]. Search by the expression "supercritical fluid extraction", more than 17,278 documents can be found (1969–2018). Between the years 2008–2018 this value is 10,831. More clarifying to the potential of SCFs, is the publication of 22,904 patents in the last 10 years (2008–2018), accounting for 68% of the total number of patents (equal to 33,703) published in the field [5]. Many countries have contributed to the development and publication of the topics regarding SCFs. In Figure 3.1 and Figure 3.2 are presented the main 15 countries which have published since 1969–2018, and 2008–2018, respectively. The values are obtained in percentage and represents in both case around 73% of the total documentation distributed by these countries [5].

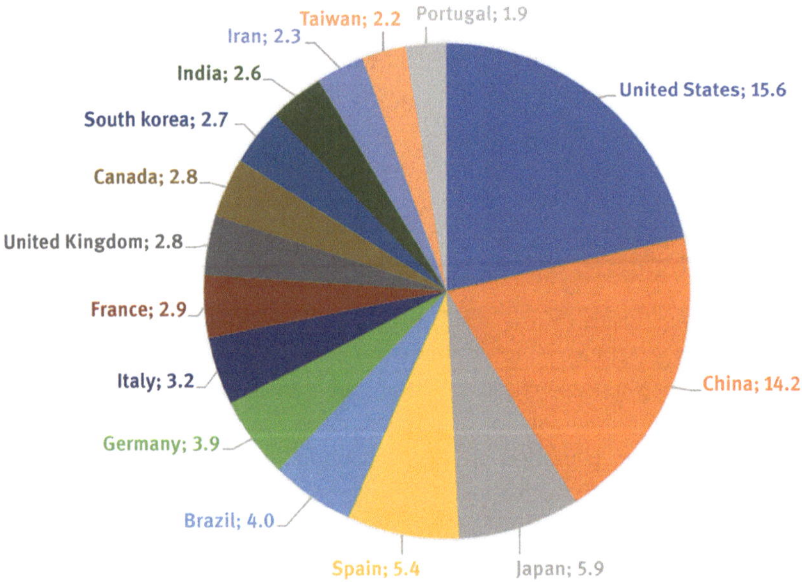

Figure 3.1: Percentage values distribution of documents produced by the firsts fifteenth countries between 1969–2018, representing around 73% of the documents.

The improvement and development with SCF have conducted to well established commercial applications. Some of the companies in the market are producers of solutions, equipment's, scale-up and optimization studies, as example: SITEC [6], JODA [7], Phasex Corporation [8], EXTRATEX [9], Eden Labs [10] and NATEX [11] while others produce simultaneous a panoply of natural botanical extracts like as: UMAX [12], Aromtech [13], Valensa [14] and Flavex Naturextrakte [15]. In Figure 3.3 two examples of industrial equipment are presented.

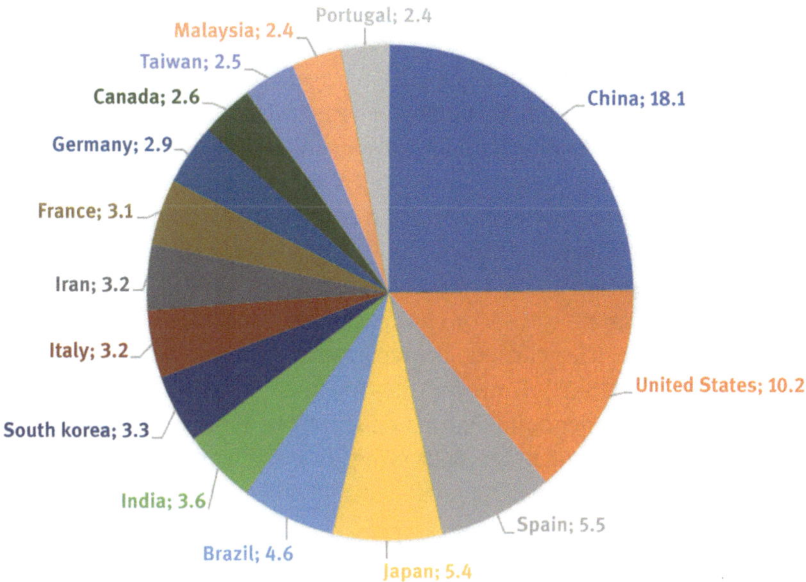

Figure 3.2: Percentage values distribution of documents produced by the firsts fifteenth countries in the last 10 years (2008–2018), representing around 73% of the documents.

Figure 3.3: Example of industrial supercritical fluid extraction plant with two extractors of 450 L. Pictures generous provided and protected by Flavex.

Many of these companies use SCFs principally SC-CO_2, applied to the processing of food materials, that has been studied since the years 1960s and denotes probably the most successful application of SCFs and related compressed fluids [16].

These extracts and oils include high-value compounds from spices, herbs, and other vegetable material, animal tissue, and microalgae which can be used as dietary supplements (also known as nutraceuticals), food and perfume ingredients personal care and fine fragrances, and, of course in pharmaceuticals.

Two important points of SCF technologies applications will be discussed: (1) A review of literature was carried out directed to microalgae as an auspicious substitute feedstock due to their high growth rate, lipid content, and photosynthetic efficiency. In addition, microalgae cultivation for fuel production does not compete with food sources and needs of arable lands and potable water [17–20]. (2) SCF technologies has established as substitute for pharmaceutical manufacturing processes to produce micronized particles carriers, coprecipitates, nanocomposite polymeric structures and liposomes [21–23]. However, these methods have still some limitations related to the control mean dimensions and size distribution and show very low encapsulation efficiency of hydrophilic drug. A novel and relatively recent method for liposome production will be focused on its potential industrial application: SuperLip.

3.2 SCFs applications to microalgae

3.2.1 Microalgae

Microalgae are unicellular photosynthetic microorganisms living in marine or freshwater environments, which convert carbon dioxide, water and light into new algal biomass. Microalgae present a complex chemical composition and the range of biochemical products make these microorganisms a very interesting resource for novel and valuable metabolites with different application.

The production of value-added compounds from microalgae is of high interest, since it can allow the consolidation of the idea of sustainable processes. Microalgae biomass has been proposed as multi-product biorefinery feed-stock [24, 25]. The algae-based biorefinery concept relies on the complete process optimization from the biomass production to the generation of different products. This includes growing and harvesting of microalgae biomass, extraction and downstream processing of value-added compounds. Microalgae biomass is composed by a high number of attractive and interesting products (polysaccharides, pigments, carotenoids, chlorophylls, vitamins, polyphenol, sterols, fatty acids, lipids, proteins and minerals), which can have numerous applications in the pharmaceutical, food, feed, cosmetic and bioenergy industries [18]. However, to get these products in the market it is necessary to have a sustainable and feasible industrial scale production.

It has been pointed out by several researchers that one of the major process constrains in microalgae biorefineries is an efficient and cost-effective extraction process [24, 25]. Bearing this in mind, the optimization of extraction processes of bioproducts from microalgae has been the main goal of several research studies and this subject has been thoroughly studied in the last decades. On the other hand, quite related to the extraction process are the pre-treatment techniques. In fact, the first operations in downstream processing of microalgae biomass, usually designated of biomass pre-treatment, such as drying and milling, can have a great impact in the extraction processes efficiency. Microalgae have a thick and hard cell wall that can hinder the extraction of internal metabolites. Therefore, most of the extraction processes of microalgae value-added are design in combination with high effective pre-treatment techniques. Otherwise, it will be very difficult to extract internal metabolites for most of the algae, since microalgae membrane and cell wall, because of its hydrocarbon composition, are very rigid and impermeable, being difficult to surpass this barrier and access the intracellular metabolites.

In this section of the chapter it will be analysed the use of supercritical fluid extraction, SFE, as a downstream process to obtain the main value-added compounds from microalgae biomass, namely, carotenoids, chlorophylls, fatty acids. Moreover, other extraction methods, required pre-treatment operations and purification technologies, as well as other alternative green extraction processes will be considered in order to achieve feasible and sustainable processes.

3.2.2 SFE to microalgae

SFE of value-added compounds from microalgae using CO_2 presents several advantages over the conventional extraction methods. It has been highlighted by several researchers the high efficiency and shortened time of extraction by SFE, as well as the higher yield that can sometimes be attained. On the other hand, the selectivity for certain compounds is more easily obtained with SFE than with conventional extraction. Moreover, for all microalgae, in a general way, the remaining biomass after lipid extraction consists mainly of protein and carbohydrates. The conventional organic solvent extraction can lead to denaturation of the proteins, unlike the SFE, what would be detrimental to its use in food or feed applications [26, 27].

Microalgae are composed of three main fractions: lipids, carbohydrates and proteins. Each of these fractions has several compounds and minor components of interest. However, extraction methods that are efficient for one fraction may not be suitable to be used with the other. In addition, minor components to be extracted can be extra or intracellular. And for the former components, as mentioned before, a first step is necessary in order to achieve the rupture of the cell wall prior to the extraction.

Microalgae biomass lipid fraction presents a great number of interesting minor compounds, such as carotenoids, fatty acids, polar lipids and chlorophylls. These compounds have been focus of attention in the last years since they are believed to play an important role in protection against a great number of chronic and acute health conditions [28], presenting powerful antioxidant properties, as well as antitumoural, anti-inflammatory and neuroprotective properties [29, 30].

SFE of lipophilic compounds from microalgae have been carried out since the final 80s [31–33] and great number of interesting studies have been published since then, as well as several reviews [34–40].

Several studies are reported in literature supporting that SFE is technical, economic and environmentally feasible and sustainable for the extraction of bioproducts from microalgae. Important microalgae such as *Botryococcus braunni*, *Chlorella vulgaris*, *Dunaliella salina*, *Haematococcus pluvialis*, *Nannochloropsis sp.*, *Neochloris*, *Scenedesmus* and *Spirulina* have been successfully submitted to SFE.

SFE using CO_2, as well as CO_2 modified with polar co-solvents (e. g. ethanol), has been demonstrated to be able to achieve very high yields of extraction, together with considerable improved selectivity and good fractionation capability [41, 42]. In addition, the lower environmental impact and the economic feasibility of SFE, as well as, the potential to scale-up the process makes SFE a very attractive technique to be used [43, 44].

In what concerns the SFE of extraction of microalgal biomass lipid components, such as carotenoids, the cell-wall disruption is a pre-treatment step of high importance to achieve a successful extraction. The cell wall disruption process should be not only effective, but also as much environmentally friendly as possible and of low cost, turning this process in a more sustainable task considering the integrated process. Among the several techniques complying with these requirements, those more suitable for the extraction of value-added compounds from microalgae combined with SFE are: milling of biomass, high-pressure homogenization (HPH), ultrasound assisted extraction (UAE) and enzymatic assisted extraction (EAE).

3.2.3 SFE to microalgae combine with other methods

The extraction of neutral lipids and fatty acids from microalga Scenedesmus sp. using subcritical CO_2 was carried out with the aim of investigating the most effective cell disruption technique for this microalga [45]. Liquid CO_2 extraction was carried out at 150 bar and 25 °C and with the addition of methanol as entrainer. The pre-treatment of the biomass was performed using several different techniques: ultrasonication, microwave radiation, gridding with liquid N_2, osmotic chock and cooling and freeze-drying. The authors verified that microwave radiation (in the presence of water) showed the highest potential to release lipids and fatty acids from the microalga biomass, and an extraction yield of 9.6% (wt.) could be reached, although Soxhlet extraction allowed to obtain a 13% yield with any pre-treatment. Moreover, these researchers also observed that freezing was the most ineffective and that, regardless the cell disruption method the obtained fatty acid methyl esters was very similar to all methods, presenting a C16-C18 composition suitable to produce biofuels.

Supercritical extraction of non-polar lipids and pigments from *Nannochloropsis gaditana* was carried out at 55 °C and 400 bar with high-pressure homogenization pre-treatment of the biomass [46] and a total extraction yield of 11.48% (wt.) was obtained. The experimental data was fitted to the Sovova's [47] broken and intact cells model. The authors verified that in the first period of the extraction rate was fast achieving a 6% yield of extraction, and that according to the AARD value the model was able to describe 99.28% of the behaviour of the experimental data.

Supercritical and subcritical fluid extraction processes assisted with ultrasounds have been recently considered of most interest for the extraction of value-added from plants and microalgae. Studies in the extraction of lutein from *Chlorella pyrenoidosa* using ultrasound-enhanced subcritical CO_2 extraction [48] have been carried out. These researchers verified that ultrasounds enhanced the subcritical CO_2 extraction of lutein using lower temperature and pressure (27 °C and 21 MPa) compared to other methods. Pressure, temperature, amount of co-solvent (ethanol) and ultrasonic power were the significant factors affecting the extraction process. The authors also performed enzymatic pre-treatment of the microalga biomass, which improved even more the extraction. Moreover, they concluded that among the most significant factors influencing the subcritical CO_2 extraction with enhanced ultrasounds and enzymatic pre-treatment, pressure was the most significant verifying that at pressures above 21 MPa the yield of extraction decreased. This was possibly due to the decrease of the diffusivity of the compounds with pressure, despite the increase of the solvent density and hence of the solvent power.

Several research works concerning ultrasound assisted extraction, UAE, of important compounds from microalgae have been also published in the last two decades. From the numerous works published in the available literature it is undisputed that UAE increases the yield of lipid extraction from microalgae when compared to conventional extraction techniques. *Nanochloropsis sp.* extraction of valuable compounds (phenols and chlorophylls) was investigated using UAE with green solvents (water, ethanol and dimethyl sulfoxide) and with mixtures of solvents (water-DMSO and water-ethanol). The extraction yield was two times higher than that of conventional water extraction and time of extraction was a critical parameter to be optimized in order to avoid degradation of chlorophylls. The selected solvent played also an important role in the extraction efficiency [49, 50] extracted lipids from *Dunaliella tertiolecta* by UAE and used factorial design to optimized the experimental parameters and evaluate the and ultrasonic power, and that the yield of extraction was 45.94% (wt.). Others significant factors were extraction time, solid/liquid ratio study, in the extraction of lipids from Chlorella vulgaris using UAE with several conventional extraction processes (Bligh and Dyer, Chen, Folch and Hara and Radin) [51]. This researcher found that the Bligh and Dyer method assisted by ultrasound resulted in the highest oil extraction yield (52.5% wt.) and that this value was higher than that obtained by conventional Soxhlet method. Moreover, in this paper the authors showed that the selection of proper solvent system is also essential to the extraction process, because it may weaken the cell wall structure, thus facilitating the cell disruption, and may allow to extract a large range of metabolites because of the different polarities. On the other hand, studies on the extraction of lutein from *Chlorella vulgaris* and economic evaluation of the process were carried out [52]. The authors used UAE with or without enzymatic pre-treatment and evaluated the effect of extraction temperature, time and solvent/solid ratio, on the yield of extraction and on the morphology of the microalgae cells. It was shown that the specific surface area of *Chlorella vulgaris* cells increased two times when UAE was carried out and eight times when UAE was performed with enzymatic pre-treatment. Also, the yield of extraction was higher for UAE with enzymatic pre-treatment, although the costs of manufacturing were the lowest when the UAE was used, suggesting that ultrasound extraction can be economically feasible process.

Nannochloropsis gaditana was evaluated in terms of its lipid extraction yield and correspondent energy consumption using UAE and methanol as solvent [53]. The yield of UAE extraction was higher than that obtained with conventional extraction process and time of extraction was reduce by half when UAE was used. Nevertheless, the energy consumption by UAE process was two to three times higher than that established as the theoretical maximum energy that can be obtained from microalgae [54].

Overall, the several researches work here detailed, as well as all the other available in literature, have shown that ultrasound assisted extraction can be a suitable alternative for the extraction of value-added compounds from microalgae. In comparison with conventional solvent extraction UAE is a more efficient and rapid method to extract the compounds from microalgae biomass, due to the strong disruption of the cell wall under ultrasonic acoustic cavitation. Also, the ultrasound extraction has no effect on the chemical structure and biological properties of the compounds. Moreover, it is suggested that ultrasound extraction can the most economical method for the extraction of compounds like carotenoids from microalgae.

3.2.4 Pressurized liquid extraction from microalgae

Microalgae lipophilic compounds can also be extracted using a green and recent technique, pressurized liquid extraction (PLE). In the recent years pressurized fluids have been using to obtain high value-added compounds from different plant matrices and microalgae without the drawbacks associated to conventional extraction processes [55–58]. Extraction using pressurize fluids has received different names, such as accelerated solvent extraction (ASE), pressurized fluid extraction (PFE), pressurized liquid extraction (PLE), pressurized hot solvent extraction (PHSE), high-pressure solvent extraction (HPSE), high-pressure, high temperature solvent extraction (HPHTSE) and subcritical solvent extraction (SSE). The use of such different names may lead to some confusion but in these it will be used the term PLE, which is the most widely accepted designation [59].

A pressurized fluid is a solvent at high temperature and pressure but always below the critical point of the fluid so that the liquid state is maintained during the whole extraction. The high pressure and temperature of the solvent induces higher solubility of the solutes, also at high temperature the fluid has low viscosity as well as surface tension, which leads to an improved penetration of the solvent in the matrix and consequently to higher mass transfer rates. All these features allow PLE to achieve higher yields with a faster extraction and a very low of solvent consumption. There are commercially available PLE equipment and these are the most used by researchers. The commercial equipment's operate in static mode, are automated, and the operational parameters that can have a profound effect on the extraction efficiency are the type of solvent used, the temperature of extraction, the time of static cycle and the number of cycles, particle size and water content of the sample [56, 58, 60, 61]. On the other hand, there are also some homemade PLE equipment, usually adapted from SFE apparatus, and these usually operate in dynamic mode [62, 63], which could improve the extraction rate by allowing a better contact between the matrix and fresh solvent pumped in a continuous way through the extraction cell. PLE employs GRAS solvents, such as ethanol and water and is suitable for a wide range of solutes, polar to non-polar [56, 58]. Other green solvents like ionic liquids [64], surfactants [58, 65], ethyl lactate [66], as well as the use of deep eutectic mixtures is also being considered as potential solvent for PLE [58, 67]. Other advantage of PLE is that at high pressure native enzymes, that can degrade some compounds, are inhibit [68]. Moreover, the PLE technique is carried out in an oxygen-free and dark-free environment, which is of high importance for bioproducts, since these are temperature-light sensible compounds.

Finally, the use of PLE in microalgae bioproducts extraction research studies can be of great interest because with this technique it is possible to work with very small amount of biomass and still have good accuracy, which can be an important advantage because the culture of microalgae can be an expensive part in microalgae research.

3.2.5 Final remarks

One of the latest extended review over the published research paper on the use of PLE to extract compounds of interest from microalgae have been published in 2015 [24]. An update of this review up to 2016 is shown in Table 3.1. The main works of PLE of bioproducts from microalgae deal with the use of static commercial systems. Microalgae, such as *Haematococcus pluvialis*, *Chlorella*, *Nannochlorpopsis*, *Scenedesmus*, *Neochloris oleoabundans* and cyanobacteria *Spirulina platensis* and *Anabaena planctonica* among other have been used in PLE studies with the aim of

Table 3.1: Review of published research works on PLE of bioproducts from microalgae and cyanobacteria.

Microalga	Bioproduct	Solvent	Yield (w%)	Refs
Chlorella vulgaris	Lipids	MeOH/$CHCl_3$	27.5	[69] Tang et al. 2016
Chlorella sorokiniana	Lipids	MeOH/$CHCl_3$	25.8	
Chlorella zofingiensis	Lipids	MeOH/$CHCl_3$	15.2	
Nanochloropsis gaditana	Lipids	MeOH/$CHCl_3$	29.8	
Spirulina platensis	lipids	ethanol	19.7	Herrero et al. [57]
Spirulina platensis	GLnA	Limonene:ethanol	1.1	[70] Golmakani et al. 2014
Anabaena planctonica	GLnA	Limonene:ethanol	0.42	
Stigeoclonium sp.	GLnA	Limonene:ethanol	0.41	
Spirulina	Lipids	Chloroform: Methanol	13.36	[71] Zheng et al. 2012
		Ethanol	n.a	
Chlorella vulgaris	Carotenoids	Acetone	n.a.	[72] Plaza et al. 2012
Chlorella vulgaris	Carotenoids	Ethanol (90%)	4.28	[73] Cha et al. 2010

obtaining lipids, essential fatty acids and carotenoids. Solvents, such as ethanol, acetone, limonene and methanol have been used and the higher yields have been obtained with ethanol (see Table 3.1 and Herrero et al. [57]). As in all extraction processes using microalgae biomass the use of milled powder increased the yield of extraction, since it increased the surface area and the solute available for the solvent. Moreover, it has also been verified in several studies that at high temperature the recovery of the compounds the was higher. Moreover, in the analysis of chlorophylls extracted by PLE it was verified that *Pheoporbidea* was not present in the extracts. Since this compound, which is a derivative form from chlorophyll a and that can cause dermatitis and therefore, its content in processed *Chlorella* is legally restricted, this result is of considerable importance.

SFE, subcritical dimethyl ether and the pressurized liquid extraction were already indicated as promising extraction techniques regarding the extraction of carotenoids from microalgae, namely those based in compressed fluids [74, 75]. With the proper optimization of these methodologies, the authors claim that it is possible to increase the amounts of carotenoids being extracted to 450 mg/g of astaxanthin recovered using ethanol as co-solvent, and at the same time to considerably reduce the extraction times (from 24–48 h to 2 h, respectively) [24]. More recent techniques appeared, namely the association of microwave extraction and the use of ionic liquids as solvents [76] which seemed to increase up to 15 times the extraction efficiency of the lipid fraction of microalgae in which carotenoids are included. Other example [48] reflected the combination of ultrasound techniques with sub-critical fluid extraction, in which the lutein yield of extraction was practically doubled (124.01 mg/100 g crude extract when compared with the Soxhlet extraction (54.64 mg/100 g crude extract).

The research in microalgae are in constant evolution and the combination of SCF with other techniques have demonstrated an evolution of the optimization of the process to improve the value of the microalgae.

3.3 SuperLip: A novel process for liposome fabrication

3.3.1 Definition of liposomes

Liposomes are spherical vesicles characterized by an inner aqueous core surrounded by one or more double layers of phospholipids [77–79]. These molecules are responsible of the backbone shape and mechanical properties of vesicles. Phospholipids are amphiphilic molecules characterized by a hydrophilic head containing different groups (such as choline in phosphatidylcholine), and a lipophilic double tail. Since they behave as surfactants, they are good candidates to form the interface between the inner aqueous compartment of liposomes and the external water bulk.

3.3.2 Use of liposomes

Liposomes are generally employed as drug carriers for the vehiculation of active molecules such as antibiotics, antioxidants, dietary supplements, vitamin and genetic material [80–83]. Lipidic vesicles are nowadays involved in several applications, depending on their high potential to preserve molecules during drug delivery. In case that the long-lasting release of an antibiotic is requested, liposome surface can be modified in order to obtain better drug retention and a prolonged delay. The versatility of these vesicles is characterized by the high biocompatibility with human cell cells. Since phospholipids are the pillars of the cell membranes, liposomes could be considered as mobile cells used as vectors to transport molecules from one tissue to another. Moreover, the cells already use a similar natural communication [84–86] system using exosomes as natural message vectors.

3.3.3 Liposomes drug release mechanisms

Drug release from liposomes can be naturally or artificially induced. In the first case, the lipidic carrier is adsorbed on the cell surface and becomes part of the cell membrane. In this manner, the drug content is directly released inside the inner core of the cell. This is due to the natural tendency of liposomes to aggregate with other systems of lipidic nature [87–90]. However, several artificial induced drug release systems have been developed in order to obtain molecule release in the time and in the site needed by the therapy. For example, pulsed drug release could be activated with the encapsulation of hollow gold nanoparticles that are sensitive to external applied magnetic field and can induce vibration and the disruption of liposome membranes. Other applicable stimuli could be the increasing of the local temperature or the variation of pH of the medium in which vesicles are suspended [91, 92].

3.3.4 Liposomes classification

Liposomes can be classified according to their mean dimensions and number of double lipidic layers, also called lamellae [93]. Single Unilamellar Liposomes are generally characterized by a mean diameter between 50 and 200 nm; whereas, Large Unilamellar Liposomes have a mean size of 200–500 nm. Finally, Giant Unilamellar Liposomes dimensions are in the range 0.5–50 µm. The observation of giant and large liposomes can be performed with optical microscope or, at least, with the confocal microscope; whereas, the observation of single unilamellar liposomes requires to work at higher magnitudes with scanning electron microscope (SEM) or transmission electron microscope (TEM). In particular, the optical and TEM microscope permit to observe the slice of liposome, with the possibility to distinguish the inner aqueous compartment from the double lipidic layer region. Also, the number of lamellae can be generally counted, one by one, exploiting this instrument. After this last analysis, it is possible to classify vesicles according to the number of lamellae, distinguishing Oligolamellar Vesicles (OLV), containing no more than 10 lamellae; MultiLamellar Vesicles (MLV), with more than 50 and MultiVesciculas Vesicles (MVV), with more liposomes disjointed included in a bigger one.

3.3.5 Liposomes methods of production

The first production method is the thin layer hydration, also called Bangham method. This conventional technique took its name from the haematologist of Brahabam institute (United Kingdom) who discovered and studied the natural liposome formation mechanism in 1965. In details, lipids are dissolved into an organic solvent in order to obtain a homogeneous solution. Then, the solvent is evaporated, and a thin layer is obtained. The third step consists in the dissolution of a hydrophilic compound in a water phase, used then for the hydration of the lipidic plane layer. After 1 h of gentle stirring, phospholipids unfold from plane configuration and liposomes are formed in this manner, in aqueous suspension.

Other fabrication processes have been developed, such as ethanol injection that is characterized by the dropwise feeding of lipidic solution in a water phase containing the drug; whereas, extrusion processes tried to obtain narrower particle size distribution of lipidic homogenized samples [94–101]. Generally, conventional methods suffer of many drawbacks such as low encapsulation efficiencies of entrapped molecules, difficult control of particle size distribution, large micrometric dimensions and high solvent residue [21, 102]. Also, batch layout used for this process is responsible of a difficult repeatability of the results. Some supercritical assisted methods have been also proposed to overcome these problems, obtaining a partial good achievement, such as the increase of encapsulation efficiency up to 50 ÷ 60% and an increased good control of particle size distributions.

3.3.6 Supercritical assisted liposome formation

To solve the problems linked to previous summarized liposomes formation techniques, a novel supercritical assisted process has been proposed. The key consists in the inversion of the traditional fabrication steps of liposomes [103–105]. First, the droplets of water are obtained and, then, they are fast surrounded by phospholipids, exploiting the high diffusion coefficient of carbon dioxide in supercritical conditions and also the jet break up of a water flow rate injected in a high pressure vessel. In details, a first feeding line of ethanol, in which phospholipids are dissolved, is pumped together with a second line of carbon dioxide. The two co-solvent are mixed in a high pressure saturator, working in the range 100 ÷ 200 bar with temperatures of 35 ÷ 45 °C. An expanded liquid is created in this manner; it transports lipids inside a formation vessel in which the same pressure and temperature have been previously fixed. A third feeding line of water in which hydrophilic compound has been dissolved is sprayed thought a nozzle in the high pressure vessel, and water droplets are created. The droplet is covered by phospholipids already fed to the system, and the drug remains confined in the inner water core. Inverted micelles are created in this manner; whereas, a second layer of phospholipids is created around the first one, falling in a water pool set at the bottom of the formation vessel. Liposomes are created and collected in liquid bulk from the bottom using an on/off valve; whereas, the expanded liquid is separated with an exit from the top.

3.3.7 Optimization of operative parameters

The variation of the operative pressure has been studied to verify its influence on vesicles formation. In particular, the increased pressure of the system resulted in the production of smaller water droplets and, as a consequence of that, smaller liposomes [103]. However, the key parameter of SuperLip process was discovered to be the Gas to Liquid Ratio of the Expanded Liquid (GLR-EL), i. e. the ratio of carbon dioxide flow rate and ethanol flow rate on mass base. In particular, working at a GLR-EL lower than 1, it was possible to obtain liposomes of micrometric mean dimensions; whereas, with GLR-EL higher than 2, nanometric dimensions were obtained for the vesicles. This meant that it was possible to control particle size distributions of the produced liposomes by changing this parameter.

As an example, a particle size distribution of a sample produced at nanometric level was reported in Figure 3.4; whereas, a scanning electron microscope image of the same sample has been provided in Figure 3.5.

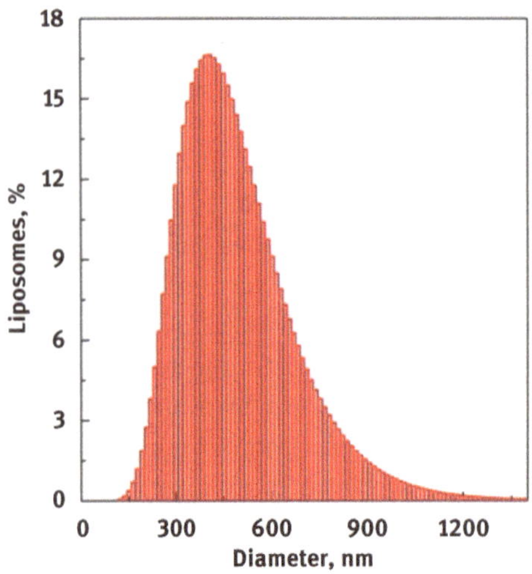

Figure 3.4: Particle size distribution of nano-liposomes produced with SuperLip.

Figure 3.5: Scanning electron microscope of nano-liposomes produced with SuperLip.

Another studied parameter was the water flow rate containing the hydrophilic compound, i. e. the third feeding line of the SuperLip process [106]. In particular, decreasing this parameter, larger water droplets were obtained and, also, larger liposomes. This fact was linked to the modification of the fluid dynamics of the system, changing the atomization region, and involving a less effective jet break up phenomenon during droplets formation.

Liposomes double lipidic layer was also improved in order to obtain a delayed and controlled drug release. In particular, cholesterol was added in the backbone of liposomes, to obtain a more compact surface and a decreased permeability of the drug carrier [107]. Also, lipidic concentration was varied, obtaining single unilamellar vesicles with lower concentrations and multilamellar vesicles with high concentrations of phospholipids.

3.3.8 SuperLip liposome-based applications

Liposomes produced with SuperLip have been proposed for several biomedical, cosmetic, nutraceutical and even textile applications, in order to demonstrate the versatility of the process and the applicability to many fields of industrial production [23, 106–108].

A cosmetic application was studied with the encapsulation of eugenol, i. e. an essential oil, either in the inner core, either in the lipidic layer of liposomes. The employment of a drug carrier better preserved the antioxidant power of the entrapped compound; in particular, the encapsulation in the inner core of liposomes resulted in the negligible reduction of the inhibition power of the antioxidants, if compared with double layer entrapment. Indeed, the inner core is considered the best compartment of encapsulation.

A nutraceutical application was also attempted for the valorization of by-products wastes such as olive pomace extract, which is characterized by tens of polyphenols exploitable for their antioxidant activity.

Several pharmaceutical applications were also studied for the encapsulation of antibiotics such as ampicillin and ofloxacin for ocular delivery, theophylline for pulmonary delivery, vancomycin and amoxicillin for antibacterial activity.

Textile field has also been recently explored; in this case, liposomes produced with SuperLip were considered for the deposition of dyes on sheep leather, in order to reduce the environmental impact of tanning process. In the sketch of Figure 3.6, a list of the SuperLip successfully tested applications has been provided.

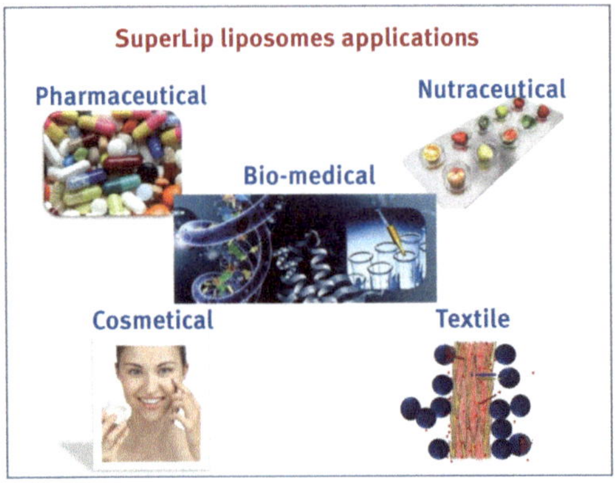

Figure 3.6: Example of liposome-based applications with SuperLip process.

3.3.9 Commercialization of the process

SuperLip process was also proposed for commercialization of liposomes-based formulations. An economic and financial analysis on the profitability of this process has been performed, providing a full Business Plan. In particular, a 5-year perspective plan was completed, in order to give a forecast of the profitability of the SuperLip process over years. As explained in the sketch, the revenues calculated for the first year are as little as about 75 k€, since a campaign of advertisement should be financed to acquire the first customers. Then, the revenues will increase up to 1.7 M€ for the last year, as expressed in the forecast for 2022 (see Figure 3.7).

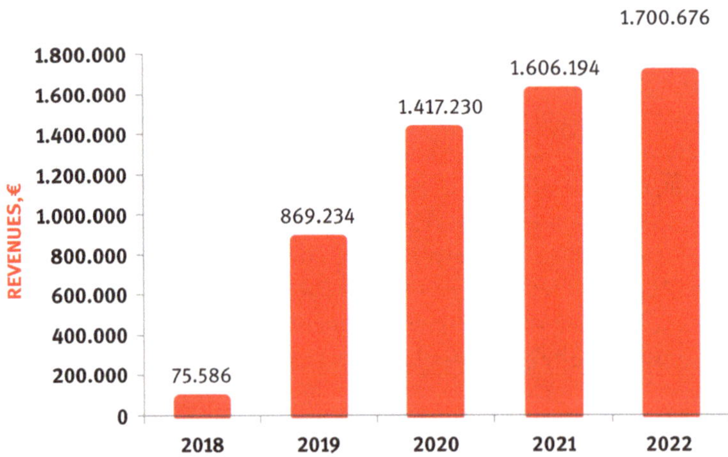

Figure 3.7: SuperLip revenues diagram for a 5-year business plan.

The business plan on SuperLip process was completed taking into account the Technology Readiness Level for this plant that was estimated to be around 6 ÷ 7, according to the European Commission guidelines. This means that the technology can work continuously, it is scalable to industrial level and that the first test sales have been already performed successfully.

3.3.10 Conclusions

SuperLip process has been successfully developed to produce liposomes at nanometric and micrometric level, with the possibility to change the order of magnitude varying the process operative parameters. Moreover, the encapsulation efficiencies of the entrapped compounds were higher that 90% in the best optimized conditions of the process. Ethanol residue in the liposomes formulations was largely lower than Food and Drug Administration imposed limits. The process was also demonstrated to be versatile and applicable in several productive fields. SuperLip was finally studied from the commercial point of view and recognized to be profitable for industrial purposes.

References

[1] Reverchon E, De Marco I. Supercritical fluid extraction and fractionation of natural matter. J Supercrit Fluids. 2006;38:146–66.
[2] Sovovà H, Stateva RP. Supercritical fluid extraction from vegetable materials. Rev Chem Eng. 2011;27:79–156.
[3] Coelho JP, Palavra AF. Supercritical fluid extraction of compounds from spices and herbs. In: Fornari T, Stateva RP, editors. High pressure fluid technology for green food processing. New York Dordrecht London: Cham Heidelberg, 2015:357–96.
[4] Stahl E, Schilz W. Extraction with supercritical gases in coupling with thin-layer chromatography - 1. Comm.: instrumentation, operation and applications. Fresenius' Zeitschrift für Anal Chemie. 1969;280:99–104.
[5] Scopus [Internet]. 2018. Available at: https://www.scopus.com/. Accessed: 15 Nov 2018.
[6] SITEC [Internet]. Available at: https://www.sitec-hp.ch/en/Willkommen. Accessed: 18 Nov 2018.
[7] JODA [Internet]. Available at: https://www.joda-tech.com/co2-scfe/. Accessed: 19 Nov 2018.
[8] Phasex [Internet]. Available at: http://www.phasex4scf.com/. Accessed: 19 Nov 2018.
[9] EXTRATEX [Internet]. Available at: http://www.extratex-sfi.com/Home. Accessed: 17 Nov 2018.
[10] Eden Labs [Internet]. Available at: https://www.edenlabs.com/. Accessed: 17 Nov 2018.
[11] NATEX [Internet]. Available at: https://www.natex.at/. Accessed: 18 Nov 2018.
[12] UMAX [Internet]. Available at: http://iumax.koreasme.com/eng/company1.html. Accessed: 18 Nov 2018.
[13] Aromtech [Internet]. Available at: http://aromtech.com/en/company/. Accessed: 18 Nov 2018.

[14] Valensa [Internet]. Available at: http://valensa.com/. Accessed: 17 Nov 2018.
[15] Flavex Naturextrakte [Internet]. Available at: http://www.flavex.com/flavex_home_cms_2_13_13.html. Accessed: 18 Nov 2018.
[16] King JW. Modern supercritical fluid technology for food applications. Annu Rev Food Sci Technol. 2014;5:215–38.
[17] Palavra AM, Coelho JP, Barroso JG, Rauter AP, Fareleira JMNA, Mainar A, et al. Supercritical carbon dioxide extraction of bioactive compounds from microalgae and volatile oils from aromatic plants. J Supercrit Fluids. 2011;60:21–7.
[18] Michalak I, Dmytryk A, Wieczorek PP, Rój E, Łęska B, Górka B, et al. Supercritical algal extracts: a source of biologically active compounds from nature. J Chem. 2015;2015:1–14.
[19] Taher H, Al-Zuhair S, Al-Marzouqi AH, Haik Y, Farid M, Tariq S. Supercritical carbon dioxide extraction of microalgae lipid: process optimization and laboratory scale-up. J Supercrit Fluids. 2014;86:57–66.
[20] Yellapu SK, Bharti , Kaur R, Kumar RL, Tiwari B, Zhang X, et al. Recent developments of downstream processing for microbial lipids and conversion to biodiesel. Bioresour Technol. 2018;256:515–28.
[21] Mozafari MR. Liposomes: an overview of manufacturing techniques. Cell Mol Biol Lett. 2005;10:711–19.
[22] Darani KK, Mozafari MR. Supercritical fluids technology in bioprocess industries: A review. J Biochem TechJ. 2009;2:144–52.
[23] Trucillo P, Campardelli R, Reverchon E. Production of liposomes loaded with antioxidants using a supercritical CO_2 assisted process. Powder Technol. 2018;323:155–62.
[24] Herrero M, Ibáñez E. Green processes and sustainability: An overview on the extraction of high added-value products from seaweeds and microalgae. J Supercrit Fluids. 2015;96:211–6.
[25] Ranjith Kumar R, Hanumantha Rao P, Arumugam M. Lipid extraction methods from microalgae: a comprehensive review. Front Energy Res. 2015;2:1–9.
[26] Stahl E, Quirin KW, Blagrove RJ. Extraction of seed oils with supercritical carbon dioxide: effect on residual proteins. J Agric Food Chem. 1984;32:938–40.
[27] Nakhost Z, Karel M, Krukonis VJ. Non-conventional approaches to food processing in CELSS. I-algal proteins; characterization and process optimization. Adv Sp Res. 1987;7:29–38.
[28] Stahl W, Sies H. Bioactivity and protective effects of natural carotenoids. Biochim Biophys Acta - Mol Basis Dis. 2005;1740:101–7.
[29] Gouveia L, Oliveira AC. Microalgae as a raw material for biofuels production. J Ind Microbiol Biotechnol. 2009;36:269–74.
[30] Brennan L, Owende P. Biofuels from microalgae-a review of technologies for production, processing, and extractions of biofuels and co-products. Renew Sustain Energy Rev. 2010;14:557–77.
[31] Choi KJ, Nakhost Z, Krukonis VJ, Karel M. Supercritical fluid extraction and characterization of lipids from Algae Scenedesmus obliquus. Food Biotechnol. 1987;1:263–81.
[32] Subra P, Boissinot P. Supercritical fluid extraction from a brown alga by stagewise pressure increase. J Chromatogr A. 1991;543:413–24.
[33] Viani M, Erazo S, Muller K, Proust P. Estudio de la biomasa y de los pigmentos carotenoides contenidos en una especie nativa de la microalga dunaliella salina sp. Rev agroquímica y Tecnol Aliment. 1989;29:539–47.
[34] Balaban M, O'Keefe S, Polak J. Supercritical fluid extraction of algae. In: King J, List G, editors. Supercritical fluid technology and lipid chemistry. Illinois: AOCS Press, 1995:247–66.
[35] Wisniak J, Korin E. Supercritical fluid extraction of lipids and other materials from algae. In: Cohen Z, Ratledge C, editors. Single cell oils. Champaign, Illinois: AOCS Press, 2005:220–38.

[36] Herrero M, Cifuentes A, Ibañez E. Sub- and supercritical fluid extraction of functional ingredients from different natural sources: plants, food-by-products, algae and microalgae - A review. Food Chem. 2006;98:136–48.
[37] Mendes R. Supercritical fluid extraction of active compounds from algae. In: Martinez J, editor. Supercritical fluid extraction of nutraceuticals and bioactive compounds. Boca Raton: CRC Press, 2007:189–213.
[38] Yen HW, Yang SC, Chen CH, Jesisca CJ. Supercritical fluid extraction of valuable compounds from microalgal biomass. Bioresour Technol. 2015;184:291–6.
[39] Harris J, Viner K, Champagne P, Jessop PG. Advances in microalgal lipid extraction for biofuel production: a review. Biofuels, Bioprod Biorefining. 2018;12:1–18.
[40] Rammuni MN, Ariyadasa TU, Nimarshana PH, Attalage RA. Comparative assessment on the extraction of carotenoids from microalgal sources: Astaxanthin from H. pluvialis and β-carotene from D. salina. Food Chem. 2019;277:128–34.
[41] Ruen-Ngam D, Shotipruk A, Pavasant P, Machmudah S, Goto M. Selective extraction of lutein from alcohol treated Chlorella vulgaris by supercritical CO2. Chem Eng Technol. 2012;35:255–60.
[42] Nobre BP, Villalobos F, Barragán BE, Oliveira AC, Batista AP, Marques PAS, et al. A biorefinery from Nannochloropsis sp. microalga - Extraction of oils and pigments. Production of biohydrogen from the leftover biomass. Bioresour Technol. 2013;135:128–36.
[43] Ferreira AF, Ribeiro LA, Batista AP, Marques PAS, Nobre BP, Palavra AMF, et al. A biorefinery from Nannochloropsis sp. microalga - Energy and CO2emission and economic analyses. Bioresour Technol. 2013;138:235–44.
[44] Crampon C, Mouahid A, Toudji SA, Le ine O, Badens E. Influence of pretreatment on supercritical CO_2 extraction from *Nannochloropsis oculata*. J Supercrit Fluids. 2013;79:337–44.
[45] Viner KJ, Champagne P, Jessop PG. Comparison of cell disruption techniques prior to lipid extraction from *Scenedesmus* sp. slurries for biodiesel production using liquid CO_2. Green Chem. 2018;20:4330–8.
[46] Sánchez-Camargo AP, Pleite N, Mendiola JA, Cifuentes A, Herrero M, Gilbert-López B, et al. Development of green extraction processes for Nannochloropsis gaditana biomass valorization. Electrophoresis. 2018;39:1875–83.
[47] Sovová H. Mathematical model for supercritical fluid extraction of natural products and extraction curve evaluation. J Supercrit Fluids. 2005;33:35–52.
[48] Fan XD, Hou Y, Huang XX, Qiu TQ, Jiang JG. Ultrasound-enhanced subcritical CO_2 extraction of lutein from *Chlorella pyrenoidosa*. J Agric Food Chem. 2015;63:4597–05.
[49] Qv XY, Zhou QF, Jiang JG. Ultrasound-enhanced and microwave-assisted extraction of lipid from *Dunaliella tertiolecta* and fatty acid profile analysis. J Sep Sci. 2014;37:2991–9.
[50] Parniakov O, Apicella E, Koubaa M, Barba FJ, Grimi N, Lebovka N, et al. Ultrasound-assisted green solvent extraction of high-added value compounds from microalgae *Nannochloropsis spp*. Bioresour Technol. 2015;198:262–7.
[51] Araujo GS, Matos LJ, Fernandes JO, Cartaxo SJM, Gonçalves LRB, Fernandes FAN, et al. Extraction of lipids from microalgae by ultrasound application: prospection of the optimal extraction method. Ultrason Sonochem. 2013;20:95–8.
[52] Deenu A, Naruenartwongsakul S, Kim SM. Optimization and economic evaluation of ultrasound extraction of lutein from *Chlorella vulgaris*. Biotechnol Bioprocess Eng. 2013;18:1151–62.
[53] Bermúdez JM, Arenillas A, Menéndez JA, Boffa L, Mantegna S, Binnelo A, et al. Optimization of microalgae oil extraction under ultrasound and microwave irradiation. J Chem Technol Biotechnol. 2014;89:1779–84.

[54] Fishman D, Majumdar R, Morello J, Pate R, Yang J. US DOE National Algal Biofuels Technology Roadmap [Internet]. US Dep. Energy, Off. Energy Effic. Renew. Energy, Biomass Program, Coll. Park. Maryl. 2010:1–124. Available at: http://biomass.energy.gov. Accessed: 20 Dec 2018.
[55] Benthin B, Danz H, Hamburger M. Pressurized liquid extraction of medicinal plants. J Chromatogr A. 1999;837:211–19.
[56] Mustafa A, Turner C. Pressurized liquid extraction as a green approach in food and herbal plants extraction: A review. Anal Chim Acta. 2011;703:8–18.
[57] Herrero M, Martín-Álvarez PJ, Señoráns FJ, Cifuentes A, Ibáñez E. Optimization of accelerated solvent extraction of antioxidants from *Spirulina platensis* microalga. Food Chem. 2005;93:417–23.
[58] Herrero M, Sánchez-Camargo AD, Cifuentes A, Ibáñez E. Plants, seaweeds, microalgae and food by-products as natural sources of functional ingredients obtained using pressurized liquid extraction and supercritical fluid extraction. TrAC - Trends Anal Chem. 2015;71:26–38.
[59] Carabias-Martínez R, Rodríguez-Gonzalo E, Revilla-Ruiz P, Hernández-Méndez J. Pressurized liquid extraction in the analysis of food and biological samples. J Chromatogr A. 2005;1089:1–17.
[60] Alonso-Salces RM, Korta E, Barranco A, Berrueta LA, Gallo B, Vicente F. Pressurized liquid extraction for the determination of polyphenols in apple. J Chromatogr A. 2001;933:37–43.
[61] Ramos L, Kristenson EM, Brinkman UA. Current use of pressurised liquid extraction and subcritical water extraction in environmental analysis. J Chromatogr A. 2002;975:3–29.
[62] Santos DT, Albuquerque CL, Meireles MA. Antioxidant dye and pigment extraction using a homemade pressurized solvent extraction system. Procedia Food Sci. 2011;1:1581–8.
[63] Mendiola JA, Herrero M, Cifuentes A, Ibañez E. Use of compressed fluids for sample preparation: Food applications. J Chromatogr A. 2007;1152:234–46.
[64] Wu H, Chen M, Fan Y, Elsebaei F, Zhu Y. Determination of rutin and quercetin in Chinese herbal medicine by ionic liquid-based pressurized liquid extraction-liquid chromatography-chemiluminescence detection. Talanta. 2012;88:222–9.
[65] Chang YQ, Tan SN, Yong JW, Ge L. Surfactant-assisted pressurized liquid extraction for determination of flavonoids from Costus speciosus by micellar electrokinetic chromatography. J Sep Sci. 2011;34:462–8.
[66] Bermejo DV, Luna P, Manic MS, Najdanovic-Visak V, Reglero G, Fornari T. Extraction of caffeine from natural matter using a bio-renewable agrochemical solvent. Food Bioprod Process. 2013;91:303–9.
[67] Pena-Pereira F, Namieśnik J. Ionic liquids and deep eutectic mixtures: Sustainable solvents for extraction processes. ChemSusChem. 2014;7:1784–800.
[68] Seabra IJ, Braga ME, Batista MT, De Sousa HC. Effect of solvent (CO2/ethanol/H2O) on the fractionated enhanced solvent extraction of anthocyanins from elderberry pomace. J Supercrit Fluids. 2010;54:145–52.
[69] Tang Y, Zhang Y, Rosenberg JN, Sharif N, Betenbaugh NJ, Wang F. Efficient lipid extraction and quantification of fatty acids from algal biomass using accelerated solvent extraction. RSC Advances. 2016;6:29127–37.
[70] Golmakani MT, Mendiola JA, Rezaei K, Ibáñez E. Pressurized limonene as an alternative bio-solvent for the extraction of lipids from marine microorganisms. J Supercrit Fluids. 2014;92:1–7.
[71] Zheng G, Guo L, Wang S, Li C, Ruo W. Purification of extracted fatty acids from the microalgae Spirulina. J Am Oil Chem Soc. 2012;89:561–66.

[72] Plaza M, Santoyo S, Jaime L, Avalo B, Cifuentes A, Reglero G, et al. Comprehensive characterization of the functional activities of pressurized liquid and ultrasound-assisted extracts from *Chlorella vulgaris*. LWT - Food Science and Technology. 2012;46:245–53.
[73] Cha KH, Lee HJ, Koo SY, Song D, Pan C. Optimization of Pressurized Liquid Extraction of Carotenoids and Chlorophylls from *Chlorella vulgaris*. J Agric Food Chem. 2010;58:793–7.
[74] Goto M, Kanda H, Wahyudiono, Machmudah S. Extraction of carotenoids and lipids from algae by supercritical CO2 and subcritical dimethyl ether. J Supercrit Fluids. 2015;96:245–51.
[75] Guedes AC, Gião MS, Matias AA, Nunes AVM, Pintado ME, Duarte CMM, et al. Supercritical fluid extraction of carotenoids and chlorophylls a, b and c, from a wild strain of *Scenedesmus obliquus* for use in food processing. J Food Eng. 2013;116:478–82.
[76] Pan J, Muppaneni T, Sun Y, Reddy HK, Fu J, Lu X, et al. Microwave-assisted extraction of lipids from microalgae using an ionic liquid solvent [BMIM][HSO4]. Fuel. 178:49–55.
[77] Bangham AD. Liposomes: Realizing their promise. Hosp Pract (Off Ed). 1992;27:51–62.
[78] Bangham AD. Liposomes: the Babraham connection. Chem Phys Lipids. 1993;64:275–85.
[79] Bangham AD. Surrogate cells or trojan horses. The discovery of liposomes. BioEssays. 1995;17:1081–8.
[80] Li R, Deng L, Cai Z, Zhang S, Wang K, Li L, et al. Liposomes coated with thiolated chitosan as drug carriers of curcumin. Mater Sci Eng C. 2017;80:156–64.
[81] Hussain A, Singh S, Sharma D, Webster TJ, Shafaat K, Faruk A. Elastic liposomes as novel carriers: Recent advances in drug delivery. Int J Nanomedicine. 2017;12:5087–108.
[82] Meissner JM, Toporkiewicz M, Matusewicz L, Machnicka B. Liposomes as non-viral carriers for genetic drugs. Postep Hig Med Dosw. 2016;70:200–9.
[83] Xu X, Burgess DJ. Liposomes as carriers for controlled drug delivery. In: Wright JC, Burgess DJ, editors. Long acting injections and implants. Advances in delivery science and technology. Boston, MA: Springer, Boston, MA, 2012:195–220.
[84] Tamm LK. Special issue on liposomes, exosomes, and virosomes. Biophys J. 2017;113: E01–E01.
[85] Antimisiaris S, Mourtas S, Papadia K. Targeted si-RNA with liposomes and exosomes (extracellular vesicles): how to unlock the potential. Int J Pharm. 2017;525:293–312.
[86] Sato YT, Umezaki K, Sawada S, Mukai SA, Sasaki Y, Harada N, et al. Engineering hybrid exosomes by membrane fusion with liposomes. Sci Rep. 2016;6:1–11.
[87] Du B, Han S, Li H, Zhao F, Su X, Cao X, et al. Multi-functional liposomes showing radiofrequency-triggered release and magnetic resonance imaging for tumor multi-mechanism therapy. Nanoscale. 2015;7:5411–26.
[88] Oerlemans C, Deckers R, Storm G, Hennink WE, Nijsen JF. Evidence for a new mechanism behind HIFU-triggered release from liposomes. J Control Release. 2013;168:327–33.
[89] Evjen TJ, Hupfeld S, Barnert S, Fossheim S, Schubert R, Brandl M. Physicochemical characterization of liposomes after ultrasound exposure - Mechanisms of drug release. J Pharm Biomed Anal. 2013;78–79C:118–22.
[90] Paasonen L, Sipilä T, Subrizi A, Laurinmäki P, Butcher SJ, Rappolt M, et al. Gold-embedded photosensitive liposomes for drug delivery: triggering mechanism and intracellular release. J Control Release. 2010;147:136–43.
[91] Xia Y, Fang M, Dong J, Xu C, Liao Z, Ning P, et al. pH sensitive liposomes delivering tariquidar and doxorubicin to overcome multidrug resistance of resistant ovarian cancer cells. Colloids Surfaces B Biointerfaces. 2018;B 570:514–20.
[92] Gouveia VM, Lopes-De-Araújo J, Costa Lima SA, Nunes C, Reis S. Hyaluronic acid-conjugated pH-sensitive liposomes for targeted delivery of prednisolone on rheumatoid arthritis therapy. Nanomedicine. 2018;13:1037–49.

[93] Yurdakul PA, EÜ T, Bölümü M, EÜ T, Bölümü M. Structure and classification of liposomes. Tekst Ve Konfeksiyon. 2007;17:243–7.
[94] Araki R, Matsuzaki T, Nakamura A, Nakatani D, Sanada S, Fu HY, et al. Development of a novel one-step production system for injectable liposomes under GMP. Pharm Dev Technol. 2018;23:602–7.
[95] Zizzari A, Bianco M, Carbone L, Perrone E, Amato F, Maruccio G, et al. Continuous-flow production of injectable liposomes via a microfluidic approach. Materials (Basel). 2017;10:1–13.
[96] Laouini A, Charcosset C, Fessi H, Holdich RG, Vladisavljevic GT. Preparation of liposomes: a novel application of microengineered membranes–from laboratory scale to large scale. Colloids Surf B Biointerfaces. 2013;112:272–8.
[97] Balbino TA, Aoki NT, Gasperini AA, Oliveira CLP, Azzoni AR, Cavalcanti LP, et al. Continuous flow production of cationic liposomes at high lipid concentration in microfluidic devices for gene delivery applications. Chem Eng J. 2013;226:423–33.
[98] Yu B, Lee RJ, Lee LJ. Microfluidic methods for production of liposomes. Methods Enzymol. 2009;465:129–41.
[99] Wagner A, Platzgummer M, Kreismayr G, Quendler H, Stiegler G, Ferko B, et al. GMP production of liposomes - A new industrial approach. J Liposome Res. 2006;16:311–19.
[100] Kukuchi H, Yamauchi H, Hirota S. A spray-drying method for mass production of liposomes. Chem Pharm Bull (Tokyo). 1991;39:1522–7.
[101] Vemuri S, Yu CD, Wangsatorntanakun V, Roosdorp N. Large-scale production of liposomes by a microfluidizer. Drug Dev Ind Pharm. 1990;16:2243–56.
[102] Maherani B, Arab-Tehrany E, Mozafari MR, Gaiani C, Linder M. Liposomes: a review of manufacturing techniques and targeting strategies. Curr Nanosci. 2011;7:436–52.
[103] Campardelli R, Espirito Santo I, Albuquerque EC, De Melo SV, Della Porta G, Reverchon E. Efficient encapsulation of proteins in submicro liposomes using a supercritical fluid assisted continuous process. J Supercrit Fluids. 2016;107:163–9.
[104] Santo IE, Campardelli R, Albuquerque EC, Vieira De Melo SA, Reverchon E, Della PG. Liposomes size engineering by combination of ethanol injection and supercritical processing. J Pharm Sci. 2015;104:3842–50.
[105] Santo IE, Campardelli R, Albuquerque EC, de Melo SV, Della Porta G, Reverchon E. Liposomes preparation using a supercritical fluid assisted continuous process. Chem Eng J. 2014;249:153–9.
[106] Campardelli R, Trucillo P, Reverchon E. Supercritical assisted process for the efficient production of liposomes containing antibiotics for ocular delivery. J CO2 Util. 2018;25:235–41.
[107] Trucillo P, Campardelli R, Reverchon E. Supercritical CO2assisted liposomes formation: optimization of the lipidic layer for an efficient hydrophilic drug loading. J CO2 Util. 2017;18:181–8.
[108] Campardelli R, Trucillo P, Reverchon E. A supercritical fluid-based process for the production of fluorescein-loaded liposomes. Ind Eng Chem Res. 2016;55:5359–65.

Venko N. Beschkov

4 Ion exchange in downstream processing in biotechnology

Abstract: Ion exchange is one of the promising methods for downstream processing in biotechnology. Its advantages are based on selectivity and therefore obtaining of products with reasonable concentration and purity, mild conditions, simple operation and saving of time and energy for product separation. Additional advantage is the possible *in situ* extraction of ionogenic products from the fermentation broth, including removal of potential inhibitors during the fermentation process.

In the case of biotechnology, ion exchange could be considered in two separate ways: ion-exchange solvent extraction and traditional liquid/solid ion exchange by ion-exchange resins. Both approaches have been studied.

In this paper, this approach is shown on two important case studies: L(+)-lactic acid and L-lysine recovery from fermentation processes.

Keywords: fermentation, product recovery, inhibitor removal, ion exchange

4.1 Introduction

As downstream processing is a key issue in biotechnology, one must also make up his mind on the advantages and the drawbacks of biotechnology. The advantage of the apparent "one-step" technologies could be erased by the long-lasting fermentation processes and the quite low product concentrations. The latter is caused either by the microbial cell metabolism or/and by the inhibition effects that some products can cause on the microbial producing strain activity.

On the contrary, usually the product recovery from the fermentation broth after the process completing is composed by series of consecutive operations, i. e. biomass separation, concentration of the fermentation broth, including evaporation or membrane processes, product solvent extraction, crystallization, drying, etc. All these processes are energy and time-consuming particularly if the product concentrations are low. Moreover, many of these separation processes are not suitable, because of their low selectivity or low yield. That is why solutions enabling the speed-up and the energy saving of the overall downstream processing scheme must be sought.

This article has previously been published in the journal Physical Sciences Reviews. Please cite as: Beschkov, V. Ion exchange in downstream processing in biotechnology *Physical Sciences Reviews* [Online] 2020, 7. DOI: 10.1515/psr-2018-0066

https://doi.org/10.1515/9783110574111-004

One promising solution of this task is the application of ion exchange for product separation during and after fermentation processes [1, 2]. For these purpose, the target product of fermentation must be electrically charged, i. e. to be cationogenic or anionogenic one. Examples of the latter are the anions of organic acids, produced by fermentation, e. g. lactic acid, citric acid, some amino acids, etc.

Ion exchange could be considered in two separate ways: the one is ion-exchange solvent extraction and the second is the traditional liquid/solid ion exchange by ion-exchange resins. Both approaches have been studied.

There are two options for ion-exchange extraction of fermentation products: product recovery after fermentation by consecutive operations: ultrafiltration with biomass recycling, ion exchange, reverse osmosis and vacuum evaporation [3], ion exchange after fermentation [3–6] or *in situ* extraction during the fermentation process [7–13]. The second one deserves more attention because it enables to extract the product directly from the broth and maintaining the optimum conditions of fermentation (pH value, low inhibitor concentration). In this paper, examples for these two options are presented. The *in situ* extraction of lactic acid as lactate anion anionogenic product will be shown. The ion-exchange extraction after the fermentation will be illustrated by the example of L-lysine.

4.2 Ion-exchange solvent extraction

In this case, the ionogenic product (e. g. organic acid anion) is extracted from the fermentation broth (an aqueous phase) by organic solvent containing active, positively charged component (usually amine) to form alkyl-ammonia salt soluble in the organic phase. Afterwards back stripping by alkaline agent takes place to recover the product as a salt, soluble in the water [14, 15] (cf. Figure 4.1).

Sometimes the process may require a preliminary concentration to make the extraction more advantageous, sometimes not. An attractive option is to apply the process of solvent extraction in parallel during the fermentation and thus to save time, energy and equipment instead of the long traditional operation scheme. However, some threats of possible poisoning the microbes by the solvent impede this approach. This threat is possible because of the solvent solubility in water, although very low. That is why the batch extraction after the end of fermentation is more acceptable.

There are different organic acids separated from fermentation broths or from model solutions. Such products are various organic acids: lactic [16–20], itaconic [21], succinic [22], tartaric [23], pyruvic [24], propionic [25], valerenic [26], mono-carboxylic [27], di-carboxylic ones [28], citric acid [29, 30], rosmarinic [31], gluconic acid [32], 2-keto-L-gluconic acid [33], etc. For these processes, the active components for extraction tri-butyl-phosphate [21, 28] and some amines are used.

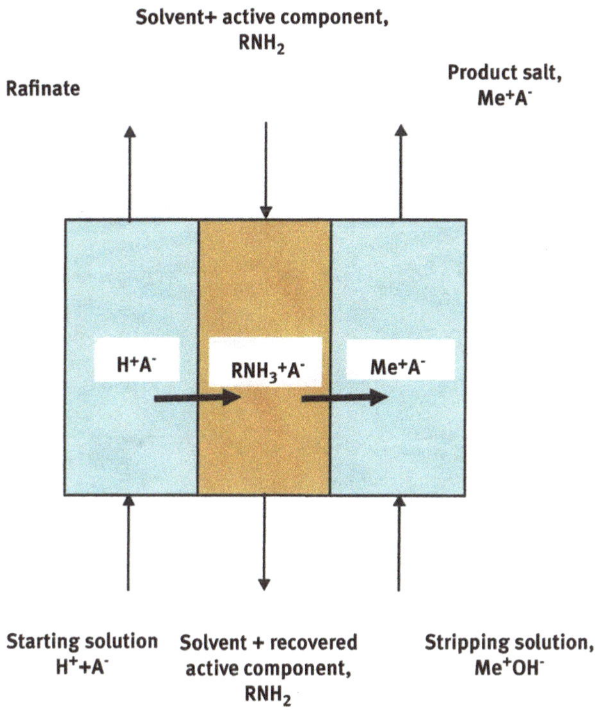

Figure 4.1: Principle scheme of continuous solvent extraction of anionic product.

Such more frequently used amines are Alamine R 336 [30, 32], Aliquat R 336 [17, 19, 21], trioctylamine [16], etc, The choice of active extractant, the organic carrier as well as the operating conditions depend on the acid strength, its concentration, the pH optimum, etc. Sometimes mixed extragents, i. e. mixtures of active components in the organic phase are applied. It was observed that in the case of lactic acid extraction the process is considerably enhanced [16, 18]. As carriers organic solvents, such as normal paraffins, chloroform, methyl isobutyl ketone or higher alcohols are used.

There is an illustration of carrier-mediated extraction of lactate with simultaneous stripping shown in Figure 4.1.

One of the main target products of such solvent extraction in biotechnology was the lactic acid, i. e. the lactate anions [16–21].

There are some results of carrier-mediated extraction of lactate with simultaneous stripping is shown in Figure 4.2.

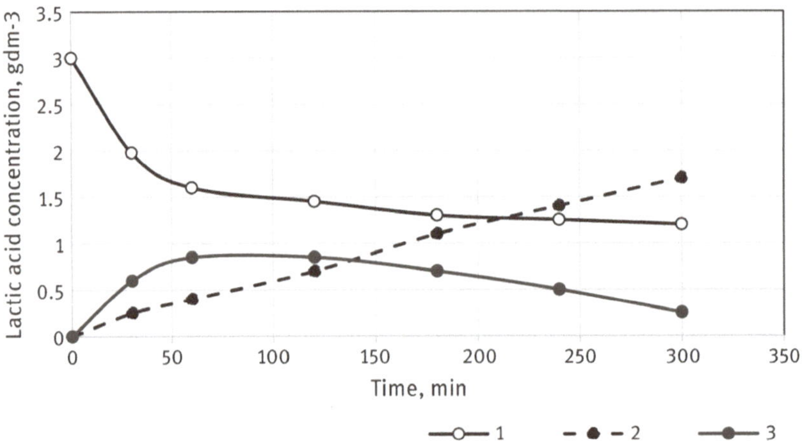

Figure 4.2: Time profile (in minutes) of lactic acid concentrations (in g dm^{-3}) in the rafinate (line 1), the stripping solution (line 2) and in the organic phase (line 3). Active component Aliquat R 336 (5% vol.) in *n*-octane. Stripping solution – 0.5 M NaCl. Data taken from [17].

4.3 Ion-exchange resins in downstream processing

4.3.1 Lactic acid extraction

Lactic acid is one of the bulk fermentation manufacturing with applications in food industry as food additive, in pharmaceutical products, as well as precursor for biodegradable polymers [34]. Its annual production reached 800,000 tons for the year 2013. It is estimated that the costs of recovery and purification in lactic acid fermentation reach about 50% of the overall production costs [8, 35]. The most important strains for lactic acid fermentation are from the genus *Lactobacillus*.

One important issue for *in situ* ion exchange is the sensitivity of microbial strains and target enzyme reactions toward the pH changes resulting from the ionogenic product accumulation. A typical example for such process and its application is the lactic acid fermentation. The latter occurs with carbohydrates (glucose, lactose) as substrates accompanied by pH drop because of the lactate accumulation:

$$C_6H_{12}O_6 \rightarrow 2CH_3CH(OH)COOH$$

with glucose as substrate at homofermentative process and

$$C_{12}H_{22}O_{11} + H_2O \rightarrow 2CH_3CH(OH)COOH \text{ with lactose.} \tag{4.1}$$

That is why no full conversion of the substrate is observed and relatively low lactate concentration is attained. It should be noted that the lactate is inhibitor of the *Lactobacillus* strains activity either as well.

One possible way to enhance the fermentation is to correct pH by some alkaline agents (sodium hydroxide, lime, limestone). However, the very lactate ions are also inhibitor for this fermentation, besides the pH change and that is why this method is not suitable in general [36].

There are some efforts to remove the product by electrodialysis [37–40], ultrafiltration [41, 42], and extraction [43–45]. An example for lactic acid extraction in hybrid bioreactor by membrane method is given in Dey and Pal [46]. It is mentioned however that electrodialysis is rather expensive process [47].

Ion exchange seems the most promising method for the simultaneous correction of pH and for product recovery [6–12]. This method is illustrated in Figure 4.3.

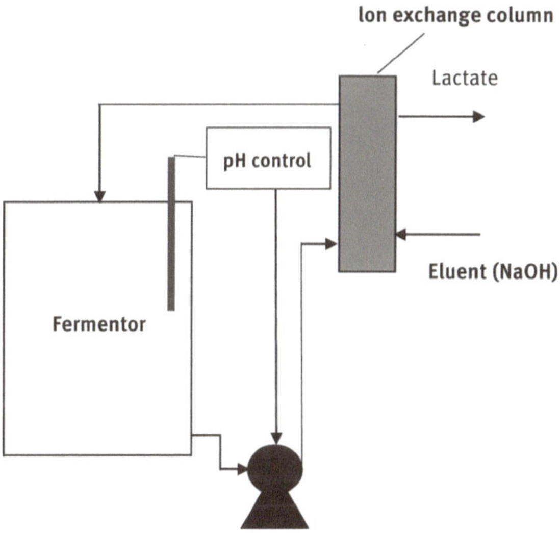

Figure 4.3: Principal sketch of *in situ* extraction from fermentation broth by ion exchange.

At the moment when pH drops below certain preset value, a pump is set on to pass the fermentation broth through a bed of anion exchange resin where lactate ions are captured, releasing hydroxyl ions to adjust pH. When the capacity of the ion-exchange resin is exhausted, regeneration by solution of sodium hydroxide is activated and the ion-exchange capacity of the resin is restored. In the same time, a solution of sodium lactate with desired concentration is obtained as a main product of the fermentation. One can see that there is a double benefit from this approach: attaining full substrate to product conversion keeping the optimum pH value and on the other hand to shorten considerably the product extraction and purification because it is recovered as lactate solution. By this method one can avoid many traditional operations in the downstream processing scheme.

It is important to select the ion-exchange resin by its selectivity toward the substrate and the ionogenic product. Higher adsorption capacity of the substrate is not desired. For the considered case [7] the adsorption capacity of two Wofatit anion-exchange resins (SL-30 and SZ-30) were tested and SL-30 showed better selectivity for lactate (cf. Figure 4.4).

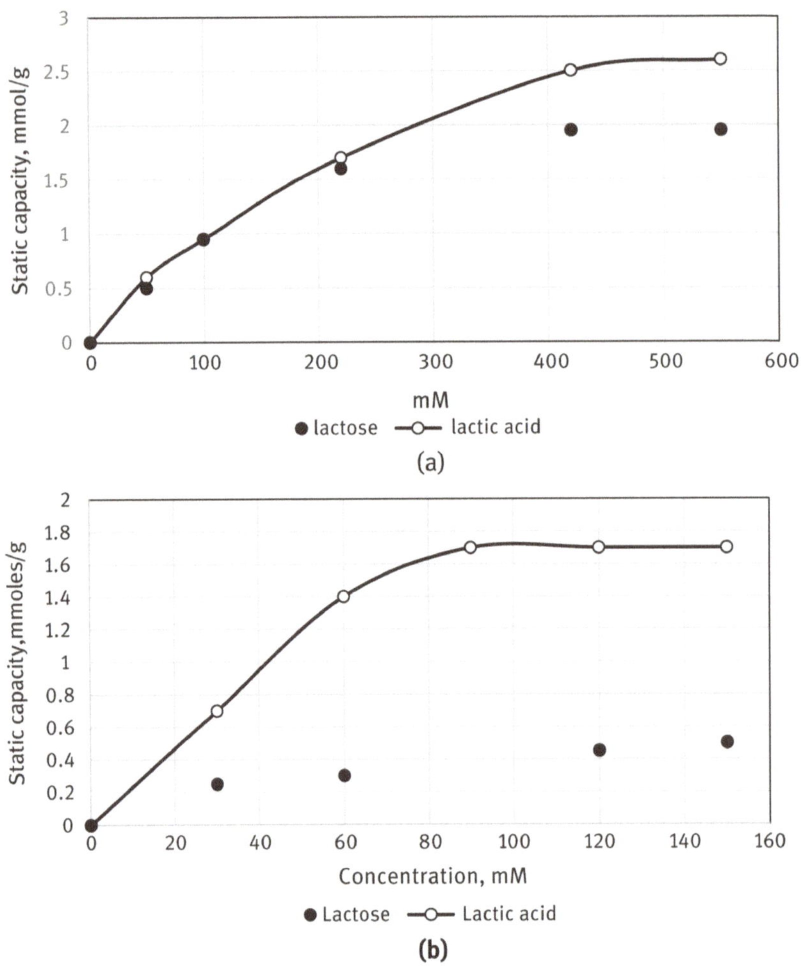

Figure 4.4: Comparison of the static capacity (mol/g.10^3 resin) vs. concentrations (mol/l.10^3) for two ion-exchange resins: Wofatit SZ-30 (a) and Wofatit SL-30 (b) Active component Aliquat R 336. Data taken from [7].

Some published results comparing the fermentation rate and the lactic acid yield for different ion-exchange resins at different initial lactose concentrations [7]. Some of the results are shown in the following Figure 4.5–Figure 4.7.

Figure 4.5 shows the comparison of control experiment with no ion exchange to one with ion-exchange resin. The lactose consumption goes more rapidly when ion exchange is applied. The same applies even stronger at higher substrate concentration

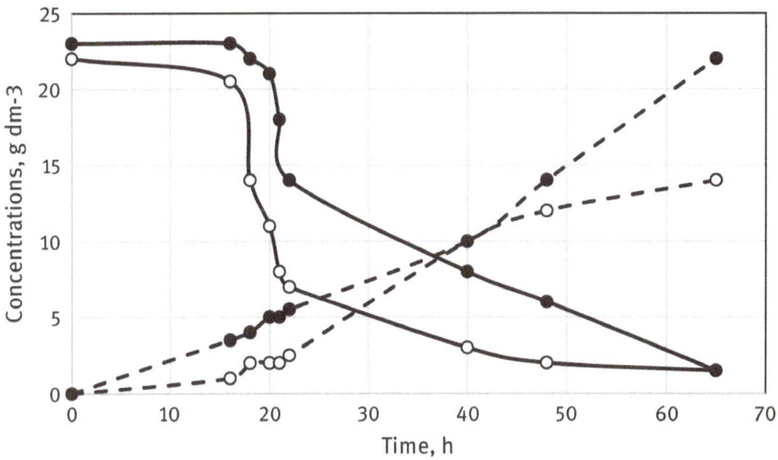

Figure 4.5: Kinetics of *in situ* lactic acid extraction from fermentation broth. Initial lactose concentration – 22 g dm^{-3}. Solid lines – lactose; dashed lines – lactic acid. Filled symbols – control experiments. Empty symbols – data with ion-exchange extraction. Ion-exchange resin Wofatit SL-30. Data taken from [7].

(cf. Figure 4.6). However, the lactic acid production rate is much higher with ion exchange at the lower concentration (i. e. 22 g/L) compared to the one at 44 g/L lactose. One possible explanation is that certain product inhibition takes place even when ion-exchange resin is applied and the lactose consumption goes faster.

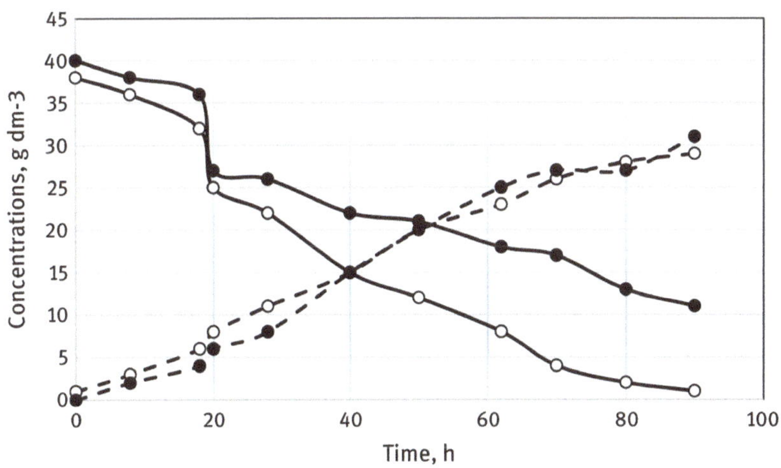

Figure 4.6: Kinetics of *in situ* lactic acid extraction from fermentation broth. Initial lactose concentration – 44 g dm^{-3}. Solid lines – lactose; dashed lines – lactic acid. Filled symbols – control experiments. Empty symbols – data with ion-exchange extraction. Ion-exchange resin Wofatit SL-30. Data taken from [7].

The fed-batch strategy is very suitable for operation. One can see that this approach gives stable results for lactose consumption and lactic acid production. Comparison of the numerical results of some reaction parameters is shown in Table 4.1.

Table 4.1: Comparison of experimental data for lactose to lactate microbial conversion with and without ion exchange by the strain *Lactobacillus casei* NBIMCC 1013. Data taken from [7].

Lactose concentration, g/l	Kinetic data	Control, batch	Ion exchange, batch	Ion exchange, fed-batch			
				Cycle 1	Cycle 2	Cycle 3	Cycle 4
22	μ, h^{-1}	0.16	0.18	0.18	0.17	0.17	0.15
	P, (g/l)/h	0.56	0.74	0.60	0.70	0.70	0.58
	η, %	96	99	99	99	99	99
55	μ, h^{-1}	0.18	0.18	–	–	–	–
	P, g/l	0.40	0.72	–	–	–	–
	η, %	60	85	–	–	–	–

It can be seen that the substrate consumption with product removal is much faster and the concentration of the inhibitor (i. e. the lactic acid) is maintained at lower values. The conversion efficiency η of substrate remains relatively constant but higher than the ones at the control experiments. This effect is considerable for the higher initial lactose concentration, i. e. 55 g/L, where the conversion efficiency is 85% compared to 60% at the control experiment.

Similar effects are observed for the production rate P. Its value is about 0.7 (g/l)/h for the experiments with ion exchange, whereas for the control experiments it is between 0.4 and 0.56 (g/l)/h.

The most important effects are observed for the case of fed-batch process (cf. Figure 4.7). As can be seen from the figure, the fed-batch process with periodical *in situ* removal of lactic acid from the fermentor leads to complete fermentation of lactose with much bigger substrate amounts (i. e. 4 × 22 g/L in the considered case). The product concentration after elution reached 60 g/L, but it can be varied depending on the amount of the elution solvent.

There is slight effect of product removal on the bacterial growth rate in the considered case, but the stationary bacterial concentration is much higher when ion exchange is applied.

A slight decrease of the ion-exchange resin capacity is observed at multiple use, possibly due to some adsorption of broth components (biostimulators, some salts) [6, 14].

There is strong interest to this process in the recent years [14, 31]. These studies, however, rely on the downstream processing after fermentation, thus omitting the

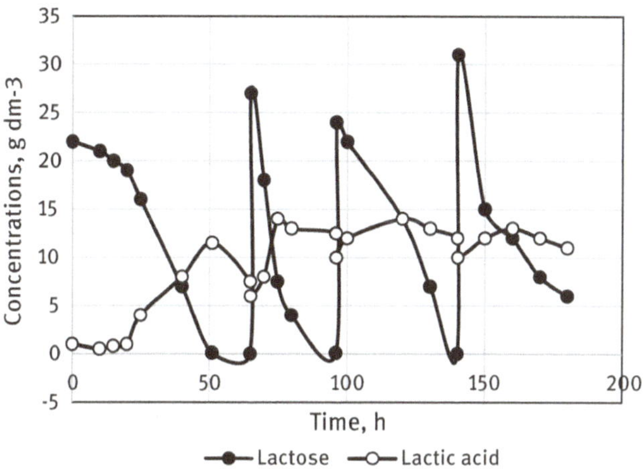

Figure 4.7: Kinetics of *in situ* lactic acid extraction from fermentation broth in a fed-batch mode. Initial lactose concentration 22 g dm^{-3}. Ion-exchange resin Wofatit SL-30. Data taken from [7].

power of the method of simultaneous fermentation and extraction the product, hence maintaining low concentration of lactate and optimum pH value in the broth.

The ion-exchange resins used for this process are mainly from Amberlite, strong ones [4–6].

4.3.2 Lysine recovery by in ion-exchange techniques

L-lysine, $HO_2CCH(NH_2)(CH_2)_4NH_2$ is an essential amino acid with application as food additive, as supplement in animal feed, etc. The global demand for L-lysine reached 1.7 kilotons in 2011 with a forecast to reach 2.5 kilotons in 2018 [48]. There are two main ways for its industrial production by fermentation from sugars or molasses by microbes of the genera *Brevibactetium* [49], *Corynebacterium* [50], *Microbacterium*, etc. [51], or by hydrolysis of proteins. Isolation of lysine from protein hydrolysates is not easy task that has been solved by precipitation as monochloride, by selective pH variation, etc.

The general downstream processing for lysine production consists of the following steps:
- Ultrafiltration for biomass and suspended solids separation;
- Ion exchange for selective separation;
- Elution with ammonia
- Evaporation and
- Crystallization.

A very important feature of all amino acids is their amphoteric nature having molecule both acid (carboxylic) and basic (amino-) groups. That is why, it is very important to select the active ionic extractant according to the amino-acid structure and its iso-electric point.

There are old-fashioned traditional methods for recovery of amino acids, and lysine in particular. They consist in precipitation as salts [52], changing the solvent, forming insoluble lysine complexes in acetone containing media (acetone) [53], etc. There are some papers, communicating for selective amino acid, including L-lysine extraction by ionic liquids [54]. Extraction with ionic liquids and two-phase aqueous systems using crown ethers is also proposed [55].

Solvent extraction with crown ethers as active component is also reported [56]. Traditional extraction using *sec*-octylphenoxy acetic acid as specialized active component in sulfonated kerosene was reported by Zhang [57, 58]. There is a group of studies on the ion-exchange extraction of lysine using liquid membrane and liquid pertraction on the principle shown in Figure 4.1. Some of them use pertraction with flat creeping film as organic phase separating two aqueous ones [59–64]. Most of them are dedicated to extraction of amino acids, namely phenylalanine [59] and lysine [60–64].

There are also some papers on electrodialysis for separation of L-lysine from protein hydrolysates [65, 66].

A sketch of the ion-exchange separation of lysine is shown in Figure 4.8. The fermentation broth containing lysine is acidified by hydrochloric acid prior to ion exchange on cationite. The latter is in an ammonia form. By this process, lysine is converted into a quaternary ammonium salt. During the ion exchange, the lysine hydrochloride is passed through the ion-exchange resin. The lysine hydrochloride cation replaces ammonia from the active site of the resin thus releasing ammonium chloride. After soaking with water, lysine is eluted by ammonium hydroxide thus regenerating the cation exchange column and releasing lysine. Then evaporation and crystallization take place to obtain the final product.

There are investigations and data on applications of ion-exchange methods for lysine separation from fermentation broth [67–70]. The main group of ion-exchange resin is from Amberlite, strong acid ones [68, 70], a Dianion SKIB [69], etc.

There is practically no modern industrial process for lysine separation without ion-exchange step.

4.3.3 Protein separation by ion-exchange chromatography [72, 73]

Protein separation is very important part of downstream processing in biotechnology. All protein products with application in medicine with specific bioactivity (e. g. vaccines) are subjected to this kind of separation and isolation.

The application of traditional separation methods common with chemical technology is strongly restricted, because of the sensitivity of the proteins toward higher temperatures, pH, ion strength, etc. That is why membrane processes (like

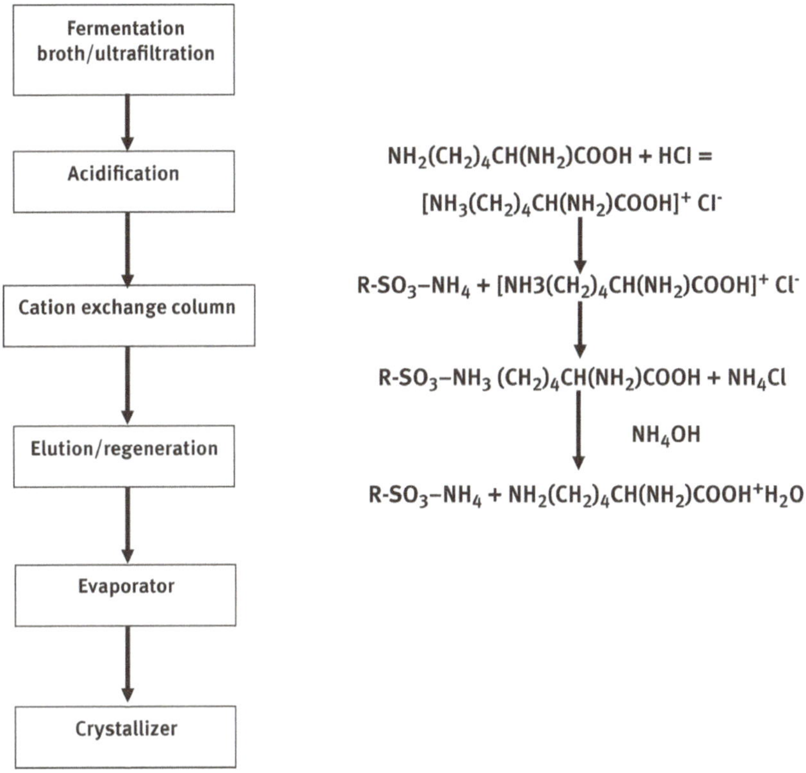

Figure 4.8: Sketch of ion-exchange separation of lysine [67].

ultrafiltration), chromatography and ion exchange are essential for isolation and purification of this group of compounds.

There are some important issues for the practical applications of ion exchange for protein separation. They depend on the chemical properties of the selected compounds. First, it is the isoelectric point (pI), at which the net charge of the protein is zero. Depending on it, an ion-exchange resin, either cationite or anionite could be selected for a certain protein. Anion exchange chromatography is applied at pH above the isoelectric point and *vice versa*. The stability of the protein depends on the pH of the environment. For a good adsorption the pH of the mobile phase must be at least one pH unit below or above of the pI value.

The interaction between the protein and ion-exchange resin depends not only on the net charge and the ionic strength, but also on the surface charge distribution of the macromolecule. It may lead to adsorption even if the net charge is zero.

The quality of protein separation depends on the selection of ion-exchange resin – strong or weak ones. As a rule strong ionites attach better weak acids and bases, whereas weak resins are suitable for strongly charged proteins.

A very important problem is the selection of eluent and to adopt certain strategy of elution. Such a common strategy is to increase the concentration of a non-buffering salt, such as sodium chloride. Its ions compete with the protein for the active site of the resin.

Elution could be attained by changing the buffer solution.

The protein separation by ion exchange and other chromatography techniques is very broad area of research and applications and it is beyond the scope of this article.

4.4 Conclusion

This short presentation has the aim to show by some practical examples the advantages of ion-exchange techniques to separate and recover some ionogenic products from fermentation broth in industrial biotechnology. The most important features of this method are its selectivity, the mild conditions (temperature and pressure), the fast accomplishment and the one-step approach in downstream processing saving equipment, time and energy for product separation. Additionally, the method is flexible to the product concentration after elution enabling saving of costs for further evaporation and crystallization.

However, this method is not quite suitable for bulky products but it is excellent option for low-scale, high-value products. Care must be taken for the choice of the resin to its product selectivity and the elution agents for each separate case.

References

[1] Zagorodni AA. Ion exchange materials: properties and applications. Amsterdam, London: Elsevier, 2007:83–90.

[2] Ion exchange chromatography. Principles and Methods, Amersham Biosciences, 10–19. ISBN 91 970490-3-4. https://www.google.bg/search?biw=1366&bih=599&ei=dHOLXKLuDtSW1fAPy7m5iAc&q=2.%09Ion+Exchange+Chromatography%2C+Amersham&oq=2.%09Ion+Exchange+Chromatography%2C+Amersham&gs_l=psy-ab.3.33i22i29i30l3.12148.12148.13172...0.0.0.96.96.1......0....1j2.gws-wiz.t769w3eeet.

[3] Gonzalez MI, Alvarez S, Riera F, Alvarez R. Economic evaluation of an integrated process for lactic acid production from ultrafiltered whey. J Food Eng. 2007;80:556–61.

[4] Bishai M, De S, Adhikari B, Banerjee R. A platform technology of recovery of lactic acid from a fermentation broth of novel substrate *Zizyphus oenophlia*. 3 Biotech. 2015;5:455–63. DOI: 10.1007/s13205-014-0240-y.

[5] Kulprathipanja S, Oroshar AR. Separation of lactic acid from fermentation broth with an anionic polymeric absorbent. US Patent 5,068,418, 1991.

[6] John RP, Nampoothiri KM, Pandey A. L(+)-Lactic acid recovery from cassava bagasse based fermented medium using anion exchange resins. Bra Arch Biol Technol. 2008;51. DOI: 10.1590/S1516-89132008000600020.

[7] Beschkov V, Peeva L, Valchanova B. Ion-exchange separation of lactic acid from fermentation broth. Hung J Ind Chem. 1995;23:135–9.

[8] Evangelista RL, Nikolov ZL. Recovery and purification of lactic acid from fermentation broth by adsorption. Appl Biochem Biotechnol. 1996;57:471–80.

[9] Kaufman EN, Cooper SP, Clement SL, Little MH. Use of a biparticle fluidized-bed bioreactor for continuous and simultaneous fermentation and purification of lactic acid. Appl Biochem Biotechnol. 1995;45/46:605–20.

[10] Davison BH, Thompson JE. Simultaneous fermentation and separation of lactic acid in a biparticle fluidized bed reactor. Appl Biochem Biotechnol. 1992;34/35:413–39.

[11] Srivastava A, Rouchoudkurt PK, Shai V. Extractive lactic acid fermentation using ion exchange. Biotechnol Bioeng. 1992;9:607–13.

[12] Yabanvar VM, Wang DI. Extractive fermentation for lactic acid production. Biotechnol Bioeng. 1991;37:1095–100.

[13] Moldes AB, Alonso JL, Parajo JC. Recovery of lactic acid from simultaneous saccharification and fermentation media using anion exchange resins. Bioproc Biosystem Eng. 2003;25:357–63.

[14] Vaccari G, Gonzalez-Vara A, Dosi E, Brigidi P, Matteuzi D. Fermentative production of L-lactic acid by Lactobacilluus casei DSM 20011 and product recovery using ion exchange resins. Appl Microbiol Biotechnol. 1993;40:23–7.

[15] Boyadzhiev L, Dimitrov K, Metcheva D. Integration of solvent extraction and liquid membrane separation: An efficient tool for recovery of bio-active substances from botanicals. Chem Eng Sci. 2006;61:4126–8.

[16] Yankov D, Molinier J, Kyuchoukov G, Albet J, Malmary G. Improvement of the lactic acid extraction. Extraction from aqueous solutions and simulated fermentation broth by means of mixed extractants and TOA, partially loaded with HCl. Chem Biochem Eng Quart. 2005;19:17–24.

[17] Lazarova Z, Peeva L, Boyanova A, Beschkov V. Recovery of the inhibitor lactic acid from fermentation broths by liquid membrane extraction. Compt Rend Acad Bulg Sci. 1993;46:65–7.

[18] Kwon YJ, Kaul R, Mattiasson B. Extractive lactic acid fermentation in poly(ethyleneimine)-based aqueous two-phase system. Biotechnol Bioeng. 1996;50:280–90.

[19] Kyuchoukov G, Marinova M, Molinier J, Albet J, Malmary G. Extraction of lactic acid by means of a mixed extractant. Ind Eng Chem Res. 2001:5635–9. Doi:10.1021/ie010137d- equilibrium, tri-3-n-octylamine.

[20] Planas J, Kozlowski A, Harris JM, Tjerneld F, Barbel HH. Novel polymer-polymer conjugates for recovery of lactic acid by aqueous two-phase extraction. Biotechnol Bioeng. 1999;66:211–18.

[21] Wasewar KL, Shende D, Keshav A. Reactive extraction of itaconic acid using tri-*n*-butyl phosphate and Aiquat 336 in sunflower oil as a non-toxic diluent. J Chem Technol Biotechnol. 2010. doi:10.1002/jctb.2500.

[22] Pratiwi AI. Study on succinic acid extraction by liquid membranes and aqueous two-phase systems containing ionic liquids, Dissertation, Doshisha University, 2014:38-50; 71–9.

[23] Marinova M, Albet J. Specific influence of the modifier (1-decanol) on the extraction of tartaric acid by different extractants. Ind Eng Chem Res. 2005;44:6534–8.

[24] Pal D, Thakre N, Kumar A, Keshav A. Reactive extraction of pyruvic acid using mixed extractants. Separation Sci Technol. 2016;51:1141–50.

[25] Wang K, Chang Z, Ma Y, Lei C, Wang J, Zhu T, et al. Study on solvent extraction of propionic acid from simulated discharged water in vitamin B12 production by anaerobic fermentation. Biores Technol. 2009;100:2878–82.

[26] Boyadzhiev L, Kancheva D, Gourdon C, Metcheva D. Extraction of valerenic acids from valerian (*Valeriana officinalis L.*) rhizomes. Pharmazie. 2004;59:727–8.

[27] Morales AF, Albet J, Kyuchoukov G, Malmary G, Monlinier J. Influence of extractant (TBP and TOA), diluent, and modifier on extraction equilibrium of monocarboxylic acids. J Chem Eng Data. 2003;48:874–86.

[28] Kyuchoukov G, Flores Morales A, Albet J, Malmary G, Molinier J. On the possibility of predicting the extraction of dicarboxylic acids with tributylphosphate dissolved in a diluent. J Chem Eng Data. 2008;53:639–47.

[29] Yordanov. B, Boyadzhiev L. Pertraction of citric acid by means of emulsion liquid membranes. J Membr Sci. 2004;238:191–7.

[30] Rodrigues Araújo EM, Bortot Coelho FE, Balarini JC, Santos Miranda TL, Salum A. Solvent extraction of citric acid with different organic phases. Adv Chem Eng Sci. 2017;7:304–24. http://www.scirp.org/journal/aces.

[31] Boyadzhiev L, Dimitrova V. Extraction and liquid membrane preconcentration of rosmarinic acid from lemon balm (*Melissa officinalis L.*) leaves. Separation Sci Technol. 2006;41:877–86.

[32] Inci I. Extraction of gluconic acid with organic solutions of alamine 336, toxic extraction of gluconic acid with organic solutions of alamine 336. Chem Biochem Eng Q. 2002;16:185–9.

[33] Blaga AC, Galaction AI, Caşcaval D. Reactive extraction of 2-keto-gluconic acid. Mechanism and influencing factors. Romanian Biotechnol Lett. 2010;15:5253–9.

[34] Wee Y-J, Kim J-N, Ryu H-W. Biotechnological production of lactic acid and its recent applications. Food Technol Biotechnol. 2006;44:163–72.

[35] Zihao W, Kefeng Z. Kinetics and mass transfer for lactic acid recovery with anion exchange method in fermentation solution. Biotechnol Bioeng. 1995;47:1–7.

[36] Li-Hong Yeh P, Bajpai K, Lanotti EL. An improved kinetic model for lactic acid fermentation. J Ferment Bioeng. 1991;71:75–7.

[37] Nomura Y, Iwahara M, Hongo M. Lactic acid production by electrodialysis fermentation using immobilized growing cells. Biotechnol Bioeng. 1987;30:788–93.

[38] Czytko M, Ishii K, Kawai K. Continuous glucose fermentationfor lactic acid production recovery of acid by electrodialysis. Chem-Ing Tech. 1987;59:952–4.

[39] Boyaval P, Corre C, Terre S. Continuous lactic acid fermentation with concentrated product recovery by ultrafiltration and electrodialysis. Biotechnol Lett. 1987;9:207–12.

[40] Siebold M, Frieling PV, Joppien R, Rindfleisch D, Schügerl K, Roper H. Comparison of the production of lactic acid by three different lactobacilli and its recovery by extraction and electrodialysis. Proc Biochem. 1995;30:81–95.

[41] Roucourt AD, Girard D, Prigent Y, Boyaval P. Continuous lactic acid fermentation with cell recycled by ultrafiltration and lactate separation by electrodialysis: model identification. Appl Envieon Microbiol. 1989;30:528–34.

[42] Russo J, Kim LJ, Hyung S. Membrane-based process for the recovery of lactic acid by fermentation of carbohydrate substrates containing sugars. US Patent 5,503,750, 1996.

[43] Schügerl K, Degener W. Recovery of low molecular weight organic components from complex aqueous mixtures by extraction. Chem Ing Tech. 1989;61:796–804.

[44] Yabanvar VM, Wang DIC. Strategies for reducing solvent toxicity in extractive fermentations. Biotechnol Bioeng. 1991;37:716–22.

[45] Lazarova Z, Peeva L. Facilitated transport of lactic acid in a stirred transfer cell. Biotechnol Bioeng. 1994;43:907–12.

[46] Dey P, Pal P. Modelling and simulation of continuous L (+) lactic acid production from sugarcane juice in membrane integrated hybrid-reactor system. Bioechem Eng J. 2013;79:15–24.

[47] Vick Roy TB, Blanch HW, Wilke CR. Microbial hollow fiber reactors. Trends Biotechnol. 1983;1:135–9. https://www.wattagnet.com/articles/15795-lysine-market-to-hit-5-9-billion-2018.

[48] Irshad S, Faisal M, Hashnu AS, Javed MM, Baber ME, Awan AR, et al. Mass production and recovery of L-lysine by microbial fermentation using Brrevibacterium flavum. J Animal Plant Sci. 2015;26:290–4.

[49] Shah AH, Hameed A, Khan GM. Fermentative production of L-Lysine: bacterial fermentation-I. J Medical Sci. 2002;2:152–7.

[50] Ezemba CC, Ozokpo CA, Anakwenze VN, Anaukwu GC, Ogbukagu CM, Ekwealor CC, et al. Lysine production of microbacterium lacticum by submerged fermentation using various hydrocarbon, sugar and nitrogen sources. Adv Microbiol. 2016;6:797–810.
[51] Rice EE. A simplified procedure for the isolation of lysine from protein hydrolysates. J Biol Chem. 1939;131:1–4.
[52] McCollum EV, Rider AA. The preparation of lysine from protein hydrolysates. J Biol Chem. 1951;190:451–3.
[53] Smirnova SV, Torocheshnikova II, Formanovsky AA, Pletnev IV. Solvent extraction of amino acids into a room temperature ionic liquid with dicyclohexano-18-crown-6. Anal Bioanal Chem. 2004;378:1369–75.
[54] Oshima T, Suetsugu A, Baba Y. Extraction and separation of a lysine-rich protein by formation of supramolecule between crown ether and protein in aqueous two-phase system. Anal Chim Acta. 2010;674:211–19.
[55] Dominguez-Perez M, Tome LIN, Freire MG, Marucho IM, Cabeza O, Coutinho JAP. Extraction of biomolecules using) aqueous biphasic systems formed by ionic liquids and aminoacids. Sep Purif Technol. 2010;75:85–91.
[56] Zhang J, Zhou X, Wang D, Zhou X, Yun F, Tan S. A comprehensive study of extraction of L-Lysine with sec-octylphenoxy-acetic acid in sulfonated kerosene. Separation Sci Technol. 2014;49:1349–56.
[57] Zhao C, Zhou X, Xiao Y, Wang D, Zhou Z, Yang Z, et al. Mechanisms of L-lysine extraction with sec-octylphenoxy-acetic acid in sulfonated kerosene. J Chem Technol Biotechnol. 2015. doi:10.1002/jctb.4799.
[58] Boyadzhiev L, Atanassova I. Extraction of phenylalanine from its dilute solutions by rotatng film pertraction. Process Biochem. 1994;29:237–43.
[59] Boyadzhiev L, Atanassova I. Recovery of L-Lysine from dilute water solutions by liquid pertraction. Biotech Bioeng. 1991;38:1059–63.
[60] Kloetzer L, Blaga AC, Postaru M, Galaction AI, Cascaval D. Selective separation of aminoacids mixture by reactive extraction and pertraction, Proc. 4th International Conference on Food Engineering and Biotechnology IPCBEE, vol.50, IACSIT Press, Singapore, 2013. v50. 14. Doi:10.7763/ipcbee
[61] Kaghazchi T, Kargari A, Yegani R, Zare A. Emulsion liquid membrane pertraction of L-lysine from dilute aqueous solutions by D2EHPA mobile carrier. Desalination. 2006;190:161–71.
[62] Thawisaeng W. L-lysine extraction from mother liquor by emulsion liquid membrane. Thesis M Eng. Chulalongkorn University, 1999.
[63] Khoshkbarchi MK, Vera JH. Reverse micellar extraction and back extraction of l-lysine with three dialkyl sodium phosphinates in pentanol/isooctane mixtures. Separation Sci Technol. 1995;30:2301–14.
[64] Sandeaux J, Sandeaux R, Gava C, Grib H, Sadat T, Belhocine D, et al. Extraction of amino acids from protein hydrolysates by electrodialysis. J Chem Technol Biotechnol. 1999. 10. 1002/(SICI)1097-4660(199803)71:3<267::AID-JCTB835>3.0.CO;2-O.
[65] Tzeng Y-M, Diosady LL, Rubin LJ. Preparation of rapeseed protein isolate by sodium hexametaphosphate extraction, ultrafiltration, diafiltration, and ion-exchange. J Food Sci. 1988. DOI: 10.1111/j.1365-2621.1988.tb09318.x. https://www.slideshare.net/kskuldeep1995/ lysine-industrial-production.
[66] Kusunose Y, Yokoyama M, Itoh H. Purification of L-lysine fermentation broth by a novel ion-exchange moving-bed method. Kagaku Kogaku Ronbunshu. 2004;30:311–17.
[67] Zurawick JJ. Separation of L-lysine by ion-exchange chromatography, Thesis, University of Tennessee, Knoxville, 1999:11–13.

[68] Melis S, Markos J, Cao G. Ion-exchange equilibria of amino acids on a strong acid resin. Ind Eng Chem Res. 1996;35:1912–20.
[69] Dimitrova A, Georgiev T, Ivanova V, Kristeva J, Ratkov A. Modelling of ion exchange step of L-lysine purification process. Biotechnol Biotechnol Equip. 1999;13:122–8.
[70] In: Cutler P. editor(s). Methods in molecular biology. vol. 244: protein purification protocols. Totowa, NJ: Humana Press, 2004:13–20. http://www.ispybio.com/search/protocols/purification%20protocol12.pdf.

Zdravka Lazarova, Venko Beschkov and Svetlozar Velizarov

5 Electro-membrane separations in biotechnology

Abstract: Membrane processes are of crucial importance for downstream processing in biotechnology. This is due to their selectivity and the mild operating conditions, enabling to extract target products without damages caused by overheating and chemical agents. Besides the most spread membrane processes like ultrafiltration and reverse osmosis, electrodialysis is very important for removal and extraction of electrically charged products, i. e. anions of organic acids, some antibiotics, etc. The electrodialysis process can be organized in batch or continuous mode. On the other hand, in the electro-crossflow filtration, the transport of target solutes across the membrane is guided by two main driving forces, the transmembrane pressure and the electric potential. This combination enables various possibilities for more selective and efficient downstream processing in biotechnology. This chapter provides a brief overview of recent achievements of electrodialysis in selected bioproducts separations and recovery. A special focus, including original experimental data, is then given to electro-filtration, which is a powerful tool creating new opportunities for performing separations on the basis of both electric charge and particle size differences.

Keywords: aminoacids, bioactive albumin, donnan dialysis, downstream processing, electric field, electrodialysis, electro-filtration, lactic acid, microfiltration, organic acids, rabbit albumin, ultrafiltration

5.1 Introduction

Membrane processes are essential for downstream processing in biotechnology, because of their selectivity and relatively mild operational conditions [1–5]. The latter are very important because of the thermal and chemical sensitivity of many bioproducts (e. g. proteins) and the high cost of biomass removal and target product(s) concentration by evaporation [6, 7]. The most spread membrane processes are ultrafiltration, reverse osmosis and electrodialysis. Whereas ultrafiltration is based on the permeation of molecules and particles with a defined size (lower than that of the membrane pores diameter) and, hence, leads to concentration of high- molecular mass products, reverse osmosis is based on the removal of the solvent (water) through a dense membrane by the application of mechanical pressure. On the other hand, Donnan dialysis is based

This article has previously been published in the journal Physical Sciences Reviews. Please cite as: Lazarova, Z., Beschkov, V., Velizarov, S. Electro-membrane separations in biotechnology *Physical Sciences Reviews* [Online] 2020, 8. DOI: 10.1515/psr-2018-0063

https://doi.org/10.1515/9783110574111-005

on the transport of counter-ions (either positively or negatively charged) from one solution (i. e. fermentation broth) to another one, passing through ion-exchange membranes with oppositely charged fixed functional groups. The way to enhance this separation process is to apply an electric field. The process is called electrodialysis. The ions contained in the feed are attracted by the electrodes situated in the two adjacent compartments until their concentrations are increased in the concentrate compartment and the feed solution is exhausted. Cations permeate across cation-exchange membranes, but are retained by anion-exchange membranes, and vice versa. Thus, ions are accumulated in alternating cells, forming a concentrate solution, while the other cells are depleted of ions, thereby forming the so-called "diluate" solution. Electrodialysis is of special interest for separation and recovery of ionogenic products in biotechnology, as organic acids are such compounds.

The electrodialysis process can be organized in batch or continuous mode.

On the other hand, membrane filtration assisted by an electric field differs from electrodialysis, which is a purely electrically driven separation process using ion exchange membranes for ion separation. In the electro-crossflow filtration, the transport of target solutes across the membrane is guided by two main driving forces, the transmembrane pressure and the electric potential, which combination creates various possibilities for more selective and efficient downstream processing in biotechnology.

The present chapter first provides a brief overview of recent achievements of electrodialysis in selected bioproducts separations and/or recovery. A special focus, including original experimental data, is then given to electro-filtration, which is a powerful tool creating new opportunities for performing bioseparations on the basis of both electric charge and particle size differences.

5.2 Examples for product recovery in biotechnology by dialysis membrane extraction

There are many examples in the literature about dialysis application in downstream processing in biotechnology. The prevalent majority with some exceptions [8, 9] is dedicated to anions extraction, e. g. organic acids recovery. The majority of them is focused on the extraction of a single carboxylic anion, but there are approaches for integrated processes allowing for simultaneous removal of more than one target ions, e. g. from fermentation broths [10–15] or hybrid downstream schemes involving with sorption processes [13].

The approach of pH control during electrodialysis for selective extraction of certain carboxylic anions from an acid mixture is of big interest [10, 14]. This process is sensitive to the acid strength, depending on the pK-values of the extracted acids and, therefore, the recovery of target acids could be tailored by maintaining or shifting the pH-value of the medium.

Selected examples of acid extractions by electrodialysis are considered below.

5.2.1 Volatile fatty acids (VFA)

This class of carboxylic acids are mainly products of anaerobic digestion of complex organic waste, like manure, straw and other lignocellulosic waste from agriculture used for the production of biogas. Their accumulation in the broth is undesirable because of the pH-shift beyond the optimum range for the methanogenic microbe metabolism in the consequent methane production. For this reason, different methods for VFA removal are considered [16, 17]. Electrodialysis has been tested as a separation process for VFA removal from fermentation broths [18–20]. Membrane separation coupled with pH gradient has been also successfully accomplished [14, 17].

Overall, separation of VFA by electrodialysis coupled with anaerobic digestion have been studied comprehensively during the last few years [19, 21–23].

5.2.2 Lactic acid

Lactic acid is one typical representative of the large-scale produced organic acids, due to its various applications, mainly as a monomer for biodegradable plastic manufacturing as well as in food industry and perfumery. Another important issue is the easy availability of relatively cheap substrates for its production, namely the whey-waste from dairy products production. The lactic acid accumulation during its production via fermentation leads to a pH drop and inhibition of the fermentation process, resulting into incomplete substrate utilization. That is why, different methods to avoid this undesirable effect have been proposed, including solvent ion exchange extraction [24], application of ion exchange resins [25–27] and electrodialysis [11, 12].

In the recent years, several studies on this process have been published. Lactic acid fermentation accompanied by electrodialysis to remove the lactic acid product-inhibitor is reported in [28, 29]. Processes involving membranes immersed in the fermentor in order to remove the lactic acid have been reported as well [30, 31].

An efficient downstream processing scheme of this fermentation including filtration, electrodialysis, ion exchange chromatography and distillation is presented in [32].

5.2.3 Aminoacids

The biotechnological manufacturing of aminoacids is a process of big importance for food industry, medicine, farmacy, animal feed, etc. The main advantage of the biotechnological way of production, in this case is the fact, that only the bioactive chiral compounds are produced. Downstream processing of aminoacids is a tricky procedure because of their amphoteric properties. One has to keep in mind that at certain pH-value (the so-called isoelectric point) the neutral molecules of each aminoacid turns into a zwitterion:

$$H_2N - R - COOH \leftrightarrow H_3N^+ - R - COO^-$$

Ion exchange and electrodialysis are the most spread treatment options in the downstream processing flow sheet [33]. There are several examples for aminoacid extraction from fermentation broth by electrodialysis in the recent years [34–37]. Those examples are for mono-sodium L-glutamate [34], methionine [35], L-phenylalanine [36] and L-threonine [37].

The selection of appropriate ion exchange membranes is limited to some adverse effects of the zwitterions on the membrane properties. That is why, dedicated membranes should be applied [38].

A number of reports are devoted to purification of aminoacids containing solutions from salts and sugars by ion exchange membranes [39–41].

Very interesting results for aminoacid separation by electrodialysis are reported by Kikuchi et al. in [42]. The application of appropriate cation-selective and anion-selective membranes enabled to separate glutamic acid, methionine and lysine from each other.

5.2.4 Inhibitor removal

Besides its major application for removal of carboxylic acids from fermentation broths [18, 24–27], electrodialysis can be used to remove other ionogenic inhibitors. A representative case is VFA removal in fermentative hydrogen production [43, 44]. The VFA separated by electrodialysis can be further converted into hydrogen as well.

There are some data on removal of inhibitors from lignocellulosic hydrolyzates by electrodialysis [45], recovery of succinic acid, which is an important biorefinery platform chemical [46–48].

5.2.5 Fuel cell applications

Other electrochemical applications of dialysis are associated with electricity generation in newly designed fuel cells [49–51]. Although there are still far from practical applications, the research efforts deserve encouragement.

5.3 Electrically enhanced crossflow membrane filtration as a separation tool in biotechnology

Attractive perspectives in the field of bio separation have been opened by coupling the cross-flow pressure-driven membrane processes with electrophoresis. The introduction of an electric field with sufficient strength and appropriate direction into the conventional dynamic filtration accelerates substantially the separation process.

In general, there is interaction between the surface charges of the membranes and the solutes in the feed solution/suspension which can lead to interference phenomena such as concentration polarisation, filter cake formation and membrane fouling at the membrane-solution interface. The combined electro-filtration process prevents to some extent the typical time-dependent permeate flux decline and improves substantially the separation selectivity. The special advantage of the electro-pressure-driven membrane processes lies in the soft anti-fouling effect; there is no need of additional shear forces to clear the membrane, it is safety to the membrane material and appropriate for treatment of sensitive bio systems. Apart from enhancement of the flux, the quality of the permeate resp. the product can be improved by applying an electric field [52].

To prevent the formation of cake on the membrane surface and membrane fouling, some specific conditions have to be fulfilled. The first one concerns the feed properties: the sign of the particles/colloids charge, pH value, feed concentration, conductivity and particle's electrophoretic mobility. The second one refers to the main process parameters (transmembrane pressure, feed flow rate, electric field strength and the type of the current flow) which have to be mutually adjusted in such way that the net particle migration is directed away from the membrane surface. In all cases, there is a critical electric field (the net particle migration velocity towards the membrane is zero) for every one of the systems to be treated which has to be determined experimentally. When the electric field strength exceeds the critical value, the electrophoretic velocity is greater than the convective flow and the concentration of dispersed materials is lowest near the membrane.

The performance of the electro-filtration process is primarily improved due to electrophoresis, the movement of charged particles or colloids relative to the liquid. In some cases, electro-osmosis (movement of the liquid relative to the stationary charged surface of the membrane or the filter cake) is found to be significant; therefore, the ζ-potential of the membrane is an important parameter too.

The electric field can be applied either across the membrane (with one electrode on either side of the membrane) or between the membrane and another electrode (the membrane itself is an electrode). The uniform distribution of the electric field along the membrane, which influences the process efficiency and the energy consumption during filtration, depends on the design of the electrodes and their location in the electro-filtration module. Wakeman and Tarleton [53] compared the performance of three module's configurations (plate, tubular and multi-tubular) and came to the conclusion that the tubular geometry offers the most effective use of electrical power, especially when the purpose is to prevent the membrane fouling.

The membranes can be made of electrically conductive or non-conductive materials. Usually, the anode is placed on the feed side because the particles are in most cases negatively charged. The best anode material is found to be titanium coated with a thin layer of a noble metal such as platinum [54]. The cathode, which could be the membrane support, is often made of stainless steel. Another alternative is the application of a membrane made of conductive material (metal or carbon) as an electrode [55]. It seems that one of the major restrictions in the commercial implementation of the electric enhanced technologies is the lack of suitable corrosion resistant and inexpensive electrode materials.

The application areas of the electro-assisted filtration include both "upstream processing" and "downstream processing" as well as the membrane reactors.

5.3.1 Applications of electro-microfiltration (EMF)

EMF of culture broth is studied by Matsumoto et al. [56] to control the selective permeation of protein solute (BSA) by using a flat MF membrane, DC (direct current), two platinum electrode plates and a cathode located on the permeate side. It was established that the feed pH value plays an important role in the separation process: at pH 7, BSA is rejected conditions (pH 3.5), BSA is charged positively and can pass the membrane.

Park [57] compared experimental results of the conventional microfiltration of haemoglobin with the cross-flow EMF. The permeate rate of EMF is found to be over 200% higher, due to the lower membrane resistance.

Hakoda et al. [58] applied an electro-microfiltration bioreactor (EMFBR) to treat a high-density culture of *Streptococcus lactis 527*, and studied the effect of the electric field on permeate flux and cell concentration. Inhibitory metabolites such as lactic acid is continuously removed from the fermentation broth during the bioprocess, whereas the cells are completely retained in the bio reactor. As a result, the cells of *S. lactis* in the EMFBR grew about 10 times as much as those of the conventional batch culture. The yield of cell against glucose and the maintenance coefficient increased substantially by application of an electric field.

5.3.2 Applications of electro-ultrafiltration (EUF)

Various aspects of the possibility for concentration of proteins by EUF have been investigated by Rios and Freund [59]. Based on results of experiments on electro-filtration of gelatine by using a tubular alumina membrane, the following general rules are stated (1) contrary to the purely pressure-driven UF, the tangential fluid velocity must be kept as low as possible; (2) there is optimal TMP that leads to a maximum permeate flux; (3) a special electric field setting technique is proposed in which the voltage has to be switch on before starting the fluid circulation.

Separation of BSA from polyethylene glycol (PEG), 20 kDa, using EUF is realised by Lentsch et al. [60]. In this case, the separation problem cannot be solved by a standard UF because BSA has about the same particle size as PEG. The BSA charge is highly dependent on the pH value of the feed solution: at pH 6.8, BSA can be repelled from the membrane whereas PEG is attracted towards the membrane by the electric forces. In this way, simultaneously transmission of PEG and retention of BSA in the retentate is reached.

To improve the performance of EUF by reducing the global membrane resistance, Mameri et al. [61] developed a module with a static deployed metal sheet as an anode which provokes turbulence near the membrane. The results show that the global hydraulic membrane resistance is reduced by half whereby the module is being more efficient for low cross-flow velocity and initial BSA-concentrations.

Oussedik et al. [62] developed a module with a static metal sheet as a turbulence promoter and an anode creating a pulsed electric field. The application of a pulsed electric field of approximately 700 V/m leads to an increase of the permeate flux by ~300%. However, the combination of a pulsed electric field and fluidized activated alumina in the feed solution (as a dynamic turbulence promoter) reduces this effect and an increase in the permeate flux of ~10% is only possible.

Zumbusch et al. [63] used alternating electrical fields as anti-fouling tool in EUF of biological suspensions. It is found that low frequency and high field strength yield the best results for electro-filtration of BSA. The effectiveness of the electric field increases with rising the conductivity up to the point where a limiting electrolytic current is reached. Moreover, increasing the protein concentration diminishes the effect of the electric field.

EUF of reverse micelles containing enzymes is studied in the first step of development of a special bioreactor for continuous enzymatic reactions [64]. It was found that the permeate flux increases with (1) increase in the AOT concentration or (2) decrease in the enzyme concentration. Both permeation flux and rejection of water increased with increase in the applied electric field strength when the cathode is installed in the permeate side. Based on these results, a new type bioreactor reactor for continuous enzymic reaction is developed.

Hakoda et al. [65] applied EUF to enzyme starch hydrolysis and found that the filtration flux was much improved (with small loss of enzyme activity) when a buffer solution was not used in the bioreactor which decreases the electrical current.

Hakoda et al. [66] used a bio reactor for lipase-catalysed hydrolysis of triolein in an AOT reversed micelle system. An UF ceramic module of tubular type separates (rejects) the Aerosol AOT reverse micelles containing lipase from the continuous iso-octane phase containing the product. The enzyme containing RMs is retained in the reactor whereas the product solution is removed from it. The application of an electric field improves the productivity of oleic acid due to the increase in the permeate flux with increasing the electric field strength.

Turkson et al. [67] performed experiments on electro filtration of BSA with a rotating module and selected four dynamic membranes made of different materials

(Zr(IV)oxide, calcium oleate, poly-2-vinylpyridine, and cadmium sulphide) as the most stable in the presence of a DC electric field for EF of BSA.

Enevoldsen et al. [68, 69] conducted crossflow EUF of five different industrial enzyme containing solutions and evaluated the process improvement in comparison with the conventional cross flow UF. For two amylases (amylase-F and amylase-S), the permeate fluxes increased 3–7 times, whereas in case of protease A and protease S, and lipase only minor flux improvement is achieved due to the low surface charge and availability of impurities. The pulsed electric field did not improve the flux. Greatest relative flux improvement is achieved at high enzyme concentration, therefore the EUF is found to be favourable as a final concentration step during UF of enzyme solutions. The conductivity is crucial for the feasibility of the EUF process, it has to be less than 2 mS/cm from an economical point of view.

UF is the most commonly-used method for the purification of whey protein. However, the required transmembrane pressure is relatively high (2–10 bar) with low limiting permeate flux. To reduce TMP and shorten filtration time without sacrificing product quality, an electro-microfiltration module is designed with Magneli Ti_4O_7/Al_2O_3 as the conductive membrane element [70]. During the EMF process, Ti_4O_7/Al_2O_3 serves as the cathode which possesses the same electrical charge as alpha-LA and beta-LG in the feed solution. The created electric field repels the whey proteins away from the membrane surface so that the components can be retained and concentrated. The designed EMF module provides a promising option for the concentration of whey protein in terms of both saved filtration time and acceptable product (retentate) quality.

5.3.3 Case study 1: Removal of BSA by MF in AC (alternating current) electric field

The present experimental study demonstrates the advantages of EMF for separation of bioactive albumin (BSA) from aqueous medium. The main purpose is to select suitable membranes and process conditions at which both permeate flux and composition remain constant over a long period of time.

A scheme of the electro-filtration set-up is shown below on Figure 5.1. The membrane module used was designed for flat sheet membranes (65 cm^2 effective surface) and two external electrodes. Due to the gas formation and other electrochemical reactions which may occur at the electrodes, external compartments are included on either side of the electrodes to avoid changes in the process streams. These compartments are separated from the retentate and permeate compartments by ion exchange membranes. The membrane module contains four chambers for the three streams: feed/retentate, permeate and electrode's rising electrolyte solution.

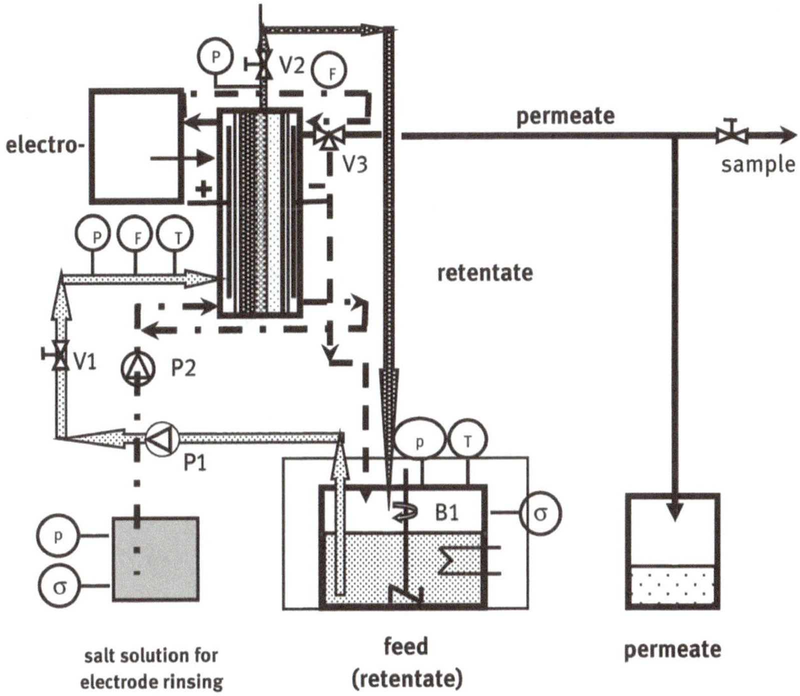

Figure 5.1: Experimental set-up for electro-filtration.

The ability of BSA to pass through different MF/UF membranes was compared using a feed containing 1 g/L BSA, feed flow rate of 300 L/h (linear velocity 2,2 m/s), and pressure at the outlet of the membrane cell of 2 bar; no current was applied. In Table 5.1, the permeate fluxes and rejection capacity of the tested membranes were compared. The *Celgard* UF-membranes (*C100 F and PS-200H*) showed the highest permeate fluxes (between ~254 and ~383 L/m^2.h); however, most of the BSA molecules were rejected by these membranes and remained in the retentate. The lowest rejection grade (~60%) showed the *Pall* MF membrane *Ultipor® NR* 0.2 µm. Therefore, this membrane was selected for the experiments in spite of the lower permeate flux.

In Figure 5.2, the effect of the alternating electric field on the permeation flux is shown. At U =0 (no current), the permeate flux decreases gradually during the first 4 h from ~ 630 to ~290 L/m^2.h. In the same time, the electrically supported flux (U = 200 V, 1 s alternating electric field) remained substantially higher (~680 L/m^2.h) and almost constant (after 2 h stationary phase for the given experimental conditions).

In Figure 5.3, the influence of the initial BSA concentration on the permeate flux in combined electro-filtration is shown. At lower protein concentration (1 g/L), there is a linear dependence between permeate flux and voltage resp. electric field strength.

Table 5.1: Permeate fluxes and BSA-rejection by different membranes.

Membrane	Time (min)	Rejection (%)	Permeate flux (L/m^2.h)
C100 F/Celgard	30	96.25	383.5
	60	95.92	364.0
UF-PS-	30	96.18	290.2
200 H/Celgard	60	96.30	254.0
Ultipor® NR/Pall	30	61.06	105.4
0,2 µm	60	60.59	107.3
Ultipor® NTG/Pall	30	85.57	118.3
0,1 µm	60	88.13	100.8
Ultipor® NT/Pall	30	86.37	94.2
0,1 µm	60	88.40	82.6

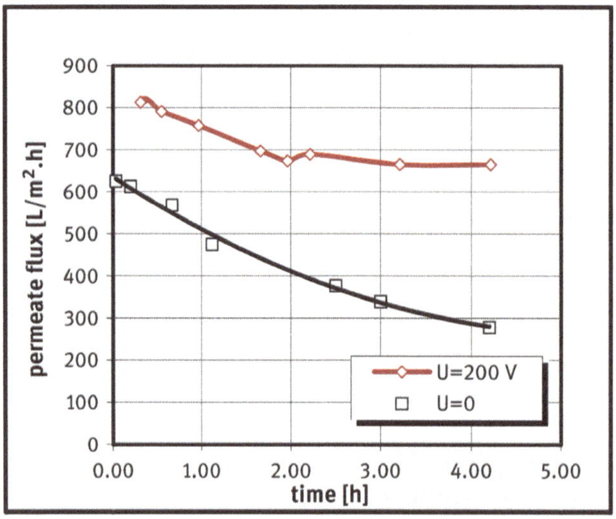

Figure 5.2: Comparison permeate fluxes at U = 200 V and U = 0.

The effect of the voltage is stronger at higher protein concentrations (5 g/L) in the feed: the permeate flux increases from 200 L/m^2.h (at U = 0 V) to ~650 L/m^2.h (at 200 V). The function permeate flux = f(voltage) becomes logarithmic one.

The rejection of BSA by the selected membrane at different voltages (in the range from 50 V to 200 V) was determined. This effect was not very pronounced for the conditions studied: BSA 1 g/L; U_F = 0.32 m/s, p_{out} = 2 bar. The rejection/retention of BSA was found to decrease slightly from 3.4% (at 50 V) to 2.9% (at 200 V). This means that more than 97% of BSA-molecules permeates the membrane into the permeate side without to affect negatively the permeate flux.

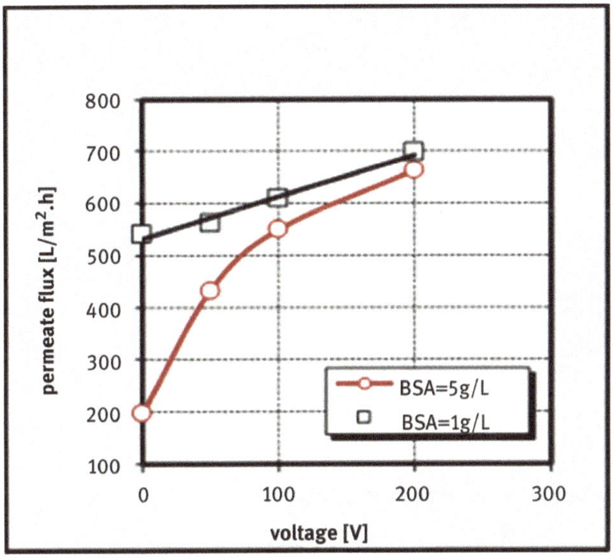

Figure 5.3: Comparison permeate fluxes at U = 0 ÷ 200 V.

5.3.4 Case study 2: EMF of rabbit albumin

The purpose of these experiments was to apply MF with an alternation electric field for treatment of a rabbit albumin solution with high conductivity (high content of ammonium sulphate). The analysis of the rabbit albumin solution by means of ZetaSizer gave important information about the particle size and ζ-potential as a function of pH.

Characterisation of suspension charges by means of the ζ-potential as a function of pH gave the isoelectric point at pH 5.35 (Figure 5.4). This means that the particles are negatively charged at pH > 5.35. The curve shows that the most appropriate pH range is the light alkaline range because the ζ-potentials, resp. the electrophoretic mobility, are several times higher at these conditions comparing to the original pH value.

The particle size was determined at three pH values: 4.08 (original solution), 2.91 and 7.27. It can be seen that at acid conditions the most of the particles have a size of 450 nm whereas at neutral pHs (7.27) most of the particles are two times smaller (Figure 5.5).

The electro-MF of the rabbit albumin solution was performed after adjustment of the pH value at pH ~ 7.2 by using an ammonium solution. The conductivity was found to be very high at 91 mS/cm. The EMF was performed at a voltage of 150 V and frequency of the pole alternation of 1 s. The electricity measured was 1.1 A (±0.2). During the conventional MF, a linear decline of the permeate flux

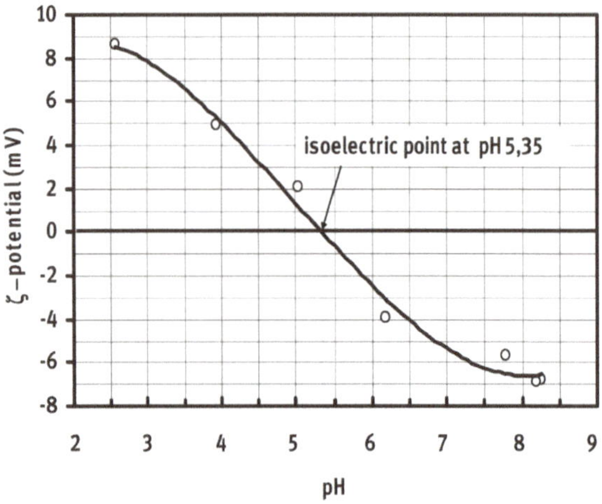

Figure 5.4: Effect of feed pH on ζ-potential of rabbit albumin solution.

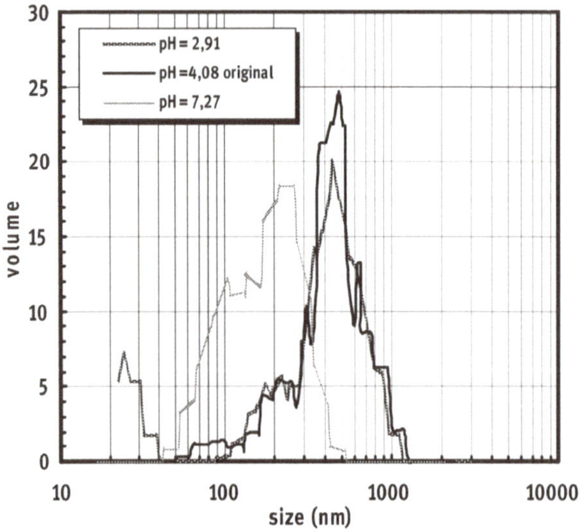

Figure 5.5: Particle size distribution as function of feed pH.

with time was observed: for 80 min, the flux decreases from 900 to 735 L/h. At the same filtration time, the EMF gave a constant flux of 950 L/m^2.h, and the rejection of rabbit albumin was measured to be below 3%. The conclusion is that EMF can be successfully applied for filtration of the bio solutions with very high conductivities.

References

[1] Jia YX, Chen X, Wang M, Wang BB. A win-win strategy for the reclamation of waste acid and conversion of organic acid by a modified electrodialysis. Sep Purif Technol. 2016;171:11–16.

[2] Kentish SE, Kloester E, Stevens GW, Scholes CA, Dumee LF. Electrodialysis in aqueous-organic mixtures. Sep Purif Rev. 2015;44:269–82.

[3] Wang J, Zheng TL, Wang QH, Wang C, Wang LH. Global trends of electrodialysis research during 1991-2014: a bibliometric analysis. J Chem Soc Pak. 2016;38:775–88.

[4] Wang XL, Wang YM, Zhang X, Jiang CX, Xu TW. Noteworthy issues for producing and transforming bioproducts by electrodalysis. J Chem Technol Biotechnol. 2014;89:1437–44.

[5] Kochendoerfer K, Gomes PS, Rutze DN, Pathapati T, De Wit P, Zaalberg M. Hybridizing simulated moving bed and electrodialysis: product purification and eluent regeneration. Chem Eng Technol. 2018;41:2418–26.

[6] Davis TA. Electrodialysis. In: Porter MC, editor. Handbook of industrial membrane technology. New Jersey: Noyes Publications, 1990.

[7] Strathmann H. Electrodialysis. In: Ho WS, Sirkar KK, editors. Membrane handbook. New York: Van Nostrand Reinhold, 1992.

[8] Miao MJ, Qiu YB, Yao L, Wu QH, Ruan HM, Van Der Bruggen B, et al. Preparation of N,N,N-trimethyl-1-adamantylammonium hydroxide with high purity via bipolar membrane electrodialysis. Sep Purif Technol. 2018;205:241–50.

[9] Shen JN, Yu J, Liu LF, Lin JY, Van Der Bruggen B. Synthesis of quaternary ammonium hydroxide from its halide salt by bipolar membrane electrodialysis (BMED): effect of molecular structure of ammonium compounds on the process performance. J Chem Technol Biotechnol. 2014;89:841–50.

[10] Arslan D, Zhang Y, Steinbusch KJ, Diels L, Hamelers HV, Buisman CJ, et al. In-situ carboxylate recovery and simultaneous pH control with tailor-configured bipolar membrane electrodialysis during continuous mixed culture fermentation. Sep Purif Technol. 2017;175:27–35.

[11] Boyaval P, Corre C, Terre S. Continuous lactic acid fermentation with concentrated product recovery by ultrafiltration and electrodialysis. Biotechnol Lett. 1987;9:207–12.

[12] Roucourt AD, Girard D, Prigent Y, Boyaval P. Continuous lactic acid fermentation with cell recycled by ultrafiltration and lactate separation by electrodialysis: model identification. Appl Environ Microbiol. 1989;30:528–34.

[13] Trinh LT, Kundu C, Lee JW, Lee HJ. An integrated detoxification process with electrodialysis and adsorption from the hemicellulose hydrolysates of yellow poplars. Bioresour Technol. 2014;161:280–7.

[14] Chandra A, Tadimeti JG, Bhuvanesh E, Pathiwada D, Chattopadhyay S. Switching selectivity of carboxylic acids and associated physico-chemical changes with pH during electrodialysis of ternary mixtures. Sep Purif Technol. 2018;193:327–44.

[15] Xu J, Su XF, Bao JW, Chen YQ, Zhang HJ, Tang L, et al. Cleaner production of citric acid by recycling its extraction wastewater treated with anaerobic digestion and electrodialysis in an integrated citric acid-methane production process. Bioresour Technol. 2015;189:186–94.

[16] Atasoy M, Owusu-Agyeman I, Plaza E, Cetecioglu Z. Bio-based volatile fatty acid production and recovery from waste streams: current status and future challenges. Bioresour Technol. 2018;268:773–86.

[17] Dai K, Wen JL, Wang YL, Wu ZG, Zhao PJ, Zhang HH, et al. Recovery of carboxylic acids produced by fermentation. Biotechnol Adv. 2014;32:873–904.

[18] Zeng RJ, Zhang F. Impacts of medium composition and applied current on recovery of volatile fatty acids during coupling of electrodialysis with an anaerobic digester. J Clean Prod. 2019;207:483–9.
[19] Scoma A, Varela-Corredor F, Bertin L, Gostoli C, Bandini S. Recovery of VFAs from anaerobic digestion of dephenolized olive mill wastewaters by electrodialysis. Sep Purif Technol. 2016;159:81–91.
[20] Pan XR, Li WW, Huang L, Liu HQ, Wang YK, Geng YK, et al. Recovery of high-concentration volatile fatty acids from wastewater using an acidogenesis-electrodialysis integrated system. Bioresour Technol. 2018;260:61–7.
[21] Outram V, Zhang Y. Solvent-free membrane extraction of volatile fatty acids from acidogenic fermentation. Bioresour Technol. 2018;270:400–8.
[22] Xu J, Guzman JJL, Andersen SJ, Rabaey K, Angenent LT. In-line and selective phase separation of medium-chain carboxylic acids using membrane electrolysis. Chem Commun. 2015a;51:6847–50.
[23] Dai K, Wen JL, Wang YL, Wu ZG, Zhao PJ, Zhang HH, et al. Impacts of medium composition and applied current on recovery of volatile fatty acids during coupling of electrodialysis with an anaerobic digester. J Clean Prod. 2019;207:483–9.
[24] Lazarova Z, Peeva L, Boyanova A, Beschkov V. Recovery of the inhibitor lactic acid from fermentation broths by liquid membrane extraction. Compt Rend Acad Bulg Sci. 1993;46:65–7.
[25] Srivastava A, Rouchoudkurt PK, Shai V. Extractive lactic acid fermentation using ion exchange. Biotechnol Bioeng. 1992;9:607–13.
[26] Beschkov V, Peeva L, Valchanova B. Ion-exchange separation of lactic acid from fermentation broth. Hung J Ind Chem. 1995;23:135–9.
[27] Pleissner D, Schneider R, Venus J, Koch T. Separation of lactic acid and recovery of salt-ions from fermentation broth. J Chem Technol Biotechnol. 2017;92:504–11.
[28] Bhanwar S, Singh A, Ganguli A. Effective conversion of industrial starch waste to L-Lactic acid by Lactococcus lactis in a dialysis sac bioreactor. Ann Microbiol. 2014;64:1447–52.
[29] Lech M, Trusek A. Batch electrodialysis of lactic acid obtained from LAB fermentation. Pol J Chem Technol. 2018;20:81–6.
[30] Fan R, Ebrahimi M, Czermak P. Anaerobic membrane bioreactor for continuous lactic acid fermentation. Membranes. 2017;7:14.
[31] Kuznetsov A, Beloded A, Derunets A, Grosheva V, Vakar L, Kozlovskiy R, et al. Biosynthesis of lactic acid in a membrane bioreactor for cleaner technology of polylactide production. Clean Technol Environ Policy. 2017;19:869–82.
[32] Neu AK, Pleissner D, Mehlmann K, Schneider R, Puerta-Quintero GI, Venus J. Fermentative utilization of coffee mucilage using Bacillus coagulans and investigation of down-stream processing of fermentation broth for optically pure L(+)-lactic acid production. Bioresour Technol. 2016;211:398–405.
[33] Buchbender F, Wiese M. Efficient concentration of an amino acid using reactive extraction coupled with bipolar electrodialysis. Chem Eng Technol. 2018;41:2298–305.
[34] Chai P, Wang JY, Lu HX. The cleaner production of monosodium L-glutamate by resin-filled electro-membrane reactor. J Memb Sci. 2015;493:549–56.
[35] Lin X, Pan JF, Zhou ML, Xu YQ, Lin JY, Shen JN, et al. Extraction of amphoteric amino acid by bipolar membrane electrodialysis: methionine acid as a case study. Ind Eng Chem Res. 2016;55:2813–20.
[36] Sun ZT, Gao XL, Zhang YS, Gao CJ. Separation and purification of L-phenylalanine from the fermentation broth by electrodialysis. Desalin Water Treat. 2016;57:22304–10.
[37] Wu J, Xu CQ, Zhang CY, Wang GS, Yan YZ, Wu CM, et al. Desalination of L-threonine (THR) fermentation broth by electrodialysis. Desalin Water Treat. 2017;81:47–58.

[38] García-García V, Montiel V, González-García J, Expósito E, Iniesta J, Bonete P, et al. The application of electrodialysis to desalting an amino acid solution. J Chem Educ. 2000;77:1477.
[39] Elisseeva TV, Shaposhnik VA, Luschik IG. Demineralization and separation of amino acids by electrodialysis with ion-exchange membranes. Desalination. 2002;149:405–9.
[40] Eliseeva TV, Tekuchev AY, Shaposhnik VA, Lushchik IG. Electrodialysis of amino acid solutions with bipolar ion-exchange membranes. Russ J Electrochem. 2001;37:423–6.
[41] Kattan Readi OM, Girones M, Nijmeijer K. Separation of complex mixtures of amino acids for biorefinery applications using electrodialysis. J Memb Sci. 2013;429:338–48.
[42] Kikuchi K, Gotoh T, Takahashi H, Higashino S, Dranoff JS. Separation of amino acids by electrodialysis with ion-exchange membranes. J Chem Eng Jpn. 1995;28:103–9.
[43] Massanet-Nicolau J, Guwy A, Dinsdale R, Premier I, Boghani H, Kaur A, et al. Biohydrogen production by fermentation and bioelectrolysis. H2FC SUPERGEN Meeting, 1-2 nd September 2016. http://www.h2fcsupergen.com/wp-content/uploads/2016/09/USW-slides-for-publication-Sept-16-part-1.pdf.
[44] Jones RJ, Massanet-Nicolau J, Guwy A, Premier GC, Dinsdale RM, Reilly M. Removal and recovery of inhibitory volatile fatty acids from mixed acid fermentations by conventional electrodialysis. Bioresour Technol. 2015;189:279–84.
[45] Lee JW, Trinh LT, Lee HJ. Removal of inhibitors from a hydrolysate of lignocellulosic biomass using electrodialysis. Sep Purif Technol. 2014;122:242–7.
[46] Fu LL, Gao XL, Yang Y, Fan AY, Hao HW, Gao CJ. Preparation of succinic acid using bipolar membrane electrodialysis. Sep Purif Technol. 2014;127:212–18.
[47] Sosa PA, Roca C, Velizarov S. Membrane assisted recovery and purification of bio-based succinic acid for improved process sustainability. J Memb Sci. 2016;501:236–47.
[48] Szczygielda M, Antczak J, Prochaska K. Separation and concentration of succinic acid from post-fermentation broth by bipolar membrane electrodialysis (EDBM). Sep Purif Technol. 2017;181:53–9.
[49] Makertihartha I, Zunita M, Rizki Z, Dharmawijaya PT. Recent advances on zeolite modification for Direct Alcohol Fuel Cells (DAFCs). In: Engineering International Conference. (Kusumawardani, R., Megawati, Subiyanto, Widjanarko, D., andKusumastuti, A. (eds)), Melville: Amer Inst Physics, vol. 1818, 2017.
[50] Zhen GY, Lu XQ, Kumar G, Bakonyi P, Xu KQ, Zhao YC. Microbial electrolysis cell platform for simultaneous waste biorefinery and clean electrofuels generation: current, situation, challenges and future perspectives. Prog Energy Combust Sci. 2017;63:119–45.
[51] Geng YK, Wang YK, Pan XR, Sheng GP. Electricity generation and in situ phosphate recovery from enhanced biological phosphorus removal sludge by electrodialysis membrane bioreactor. Bioresour Technol. 2018;247:471–6.
[52] Huotari HM, Trägårth G, Huisman IH. Crossflow membrane filtration enhanced by an external DC electric field: A review. TransIChem E. 1999;77, Part A:461–8.
[53] Wakeman RJ, Tarleton ES. Membrane fouling prevention in cross flow microfiltration. Chem Eng Sci. 1987;42:829–42.
[54] Bowen WR. Electrochemical aspects of microfiltration and ultrafiltration. In: Howell JA, Sanchez V, Field RW, editors. Membranes in bioprocessing. Cambridge: Chapman & Hall, 1993:265–92.
[55] Huotari HM, Huisman IH, Trägårth G. Electrically enhanced crossflow membrane filtration of oily wastewaters using the membrane as a cathode. J Membrane Sci. 1998;156:49–60.
[56] Matsumoto Y, Sudoh M, Suzuki Y. Selective permeation of protein solute from culture breath by electro-microfiltration. Kagaku Kogaku Ronbunshu. 1998;24:810–13.

[57] Park YG. Improvement of dead-end filtration during crossflow electro-microfiltration of proteins. J Ind Eng Chem. 2005;11:692–9.
[58] Hakoda M, Nakamura K. High density culture of streptococcus lactis 527 using electro-microfiltration Bioreactor. Kagaku Kogaku Ronbunshu. 1996;22:590–6.
[59] Rios GM, Freund P. Design and performance of ceramic EUF process for protein concentration. Key Eng Mater. 1991;61–62 (Inorg. Membr., ICIM-91):255–60.
[60] Lentsch S, Imar P, Orozco JL. Enhanced separation of albumin-poly(ethyleneglycol) by combination of ultrafiltration and electrophoresis. J Membrane Sci. 1993;80:221–32.
[61] Mameri N, Oussedik S, Yeddou R, Piron DL, Belhocine D, Lounici H, et al. Enhancement of ultrafiltration flux by coupling static turbulence promoter and electric field. Sep Purif Technol. 1999;17:203–11.
[62] Oussedik S, Belhocine D, Grib H, Lounici H, Piron DL, Mameri N. Enhanced ultrafiltration of bovine serum albumin with pulsed electric field and fluidized activated alumina. Desalination. 2000;127:59–68.
[63] Zumbusch VP, Kulcke W, Brunner G. Use of alternating electrical fields as anti-fouling strategy in ultrafiltration of biological suspensions – introduction of a new experimental procedure for crossflow filtration. J Membr Sci. 1998;142:75–86).
[64] Nakamura K, Hakoda M, Electro-ultra filtration bioreactor for enzymatic reactionsin reverse micelles, Biochem.Eng. for 2001: Proc. Asia-Pacific Biochem.Eng.Conf. 1992, 433-437, Eds. S. Furusaki, I.Endo, R.Matsuumo, Springer Verglag Tokio.
[65] Hakoda M, Chiba T, Nakamura K. Characteristics of electro-ultrafiltration bioreactor. Kagaku Kogaku Ronbunshu. 1991;17:470–6.
[66] Hakoda M, Enomoto A, Nakamura K. High densitiy culture of Streptococcus lactis 527 using electro-microfiltration bioreactor. Kagaku Kogaku Ronbunshu. 1996;22:590–6.
[67] Turkson AK, Mikhlin JA, Weber ME. Dynamic membranes. I.Determination of optimum formation conditions and electro filtration of bovine serum albumin with a rotating module. Sep Sci Technol. 1989;24:1261–91.
[68] Enevoldsen AD, Hansen EB, Jonsson G. Electro-ultrafiltration of industrial enzyme solutions. J Memb Sci. 2007;299:28–37.
[69] Enevoldsen AD, Hansen EB, Jonsson G. Electro-ultrafiltration of amylase enzymes: process design and economy. Chem Eng Sci. 2007;62:6716–25.
[70] Geng P, Chen G. Electro-microfiltration concentration of whey protein using Magneli titanium sub-oxide modified ceramic membrane. In: Asia Pacific Conf. of Chem. Eng. Congress 2015: APCChE 2015, incorp. CHEMECA 2015, Melbourne, Eng. Australia, 27.09-1.10, 2015 :641.

Dragomir Yankov

6 Aqueous two-phase systems as a tool for bioseparation – emphasis on organic acids

Abstract: Aqueous two-phase systems (ATPS) are universally recognized as an excellent alternative to the conventional separation techniques in the biotechnology, because of their undoubted advantages such as mild and biocompatible conditions, high water content, low interfacial tension, ease of process integration and scale up, etc. The formation of ATPS is due to the incompatibility of two polymers in a common solution. Other types of ATPS are formed by polymer/salt, ionic and/or non-ionic surfactants, inorganic salt/short-chain alcohols, and based on room temperature ionic liquids. ATPS are successfully used (even in large scale) for cells, enzyme and protein separation, while their application for recovery of small molecules such as organic acids, antibiotics, alcohols is more complicated as they are usually hydrophilic and tend to distribute evenly between the phases. The purpose of this paper is to overview and summarize the efforts made for the application of different types of ATPS for the separation of organic acids.

Keywords: Aqueous two-phase systems, organic acids, separation

Aqueous two-phase systems (ATPSs) were firstly noted in 1896 by Beijerinck [1] who described the “incompatibility” of starch and agar solutions. Later on, Albertsson [2] used different ATPS for separation of cells and macromolecules. ATPS are formed when two water-soluble components (two polymer solutions, a polymer solution with kosmotropic salt or a chaotropic and a kosmotropic salt) are mixed in appropriate concentrations or at a definite temperature. In recent years, the application of ATPSs for separation of various biosubstances constantly increases. This is due to the fact that such systems offer very mild conditions for separation, that are not harmful to delicate biomolecules and do not lead to loss of activity of enzymes, antibiotics, etc. Other advantages of ATPS are: high separation yield, fast phase separation, effortless scale-up, high biocompatibility. Recently, new types of ATPS have been investigated – inorganic salt/short-chain alcohols, ionic and/or non-ionic surfactants and ionic liquids based on room temperature.

Figure 6.1 represents phase diagram of an ATPS system.

This article has previously been published in the journal Physical Sciences Reviews. Please cite as: Yankov, D. Aqueous two-phase systems as a tool for bioseparation – emphasis on organic acids *Physical Sciences Reviews* [Online] 2020, 9. DOI: 10.1515/psr-2018-0067

https://doi.org/10.1515/9783110574111-006

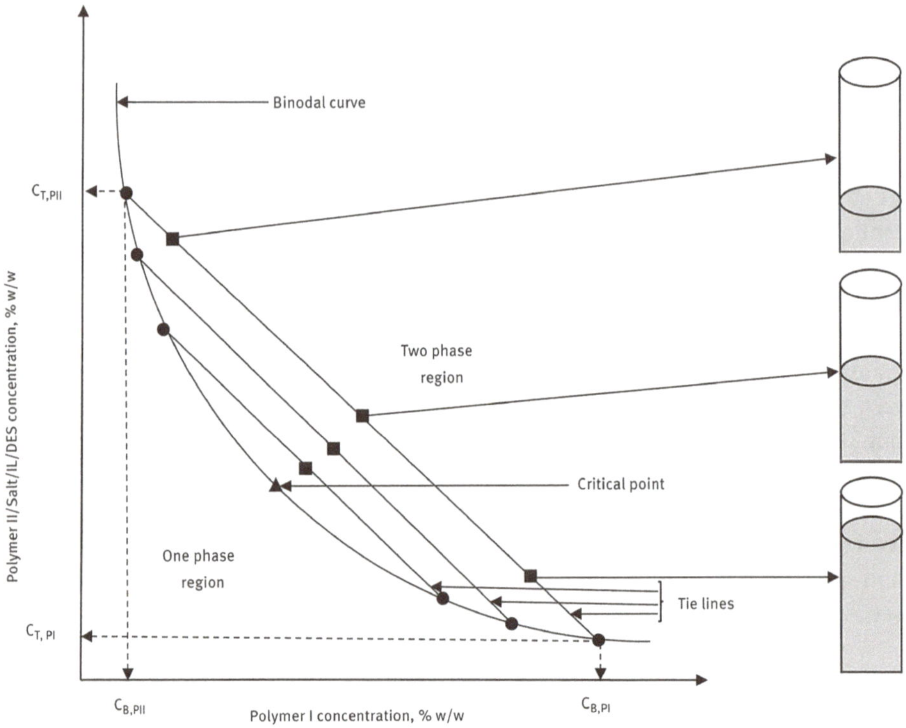

Figure 6.1: A sketch of phase diagram of an ATPS. ● – binodal points composition; ■ – tie-lines points composition; ▲ – critical point.

Despite the extensive research on properties and applications of ATPS up to now, there is not a sophisticated theory capable to describe and predict the formation of two-phase systems and the partition of target biomolecules. Numerous factors affect ATPS formation and phase separation – pH, temperature, molecular weight of the polymers, type and concentration of added salts. The properties of target substances – weight, charge, and hydrophobicity – also might influence their distribution behavior. In general, relatively large objects – cells, cell's debris, and organelles, proteins, and enzymes are distributed predominantly in one of the phases, while the low molar mass components are distributed equally between the phases. In view to increasing the partition of the desired component between the phases an affinity ligand is introduced into the system, either as a free ligand or attached to the polymer or even to the target substance. The reader can refer to some excellent reviews published recently for more information about the properties of different ATPS and their uses for biomolecules separation [3–8]. Despite the tremendous progress in understanding the separation of biomolecules in ATPS, it is still more art than science.

The aim of this study is to summarize the latest achievements in the application of various types of ATPS in the separation and concentration of low-molecular-weight organic acids produced by fermentation. The amino acids are intentionally excluded from this review because they are subject of so many investigations that deserve a separate consideration. With regard to antibiotics, only some papers (mainly on clavulanic and 6-aminopenicillanic acid) are included as illustrative examples.

Possible applications of ATPS in bioproducts separation are given in Figure 6.2.

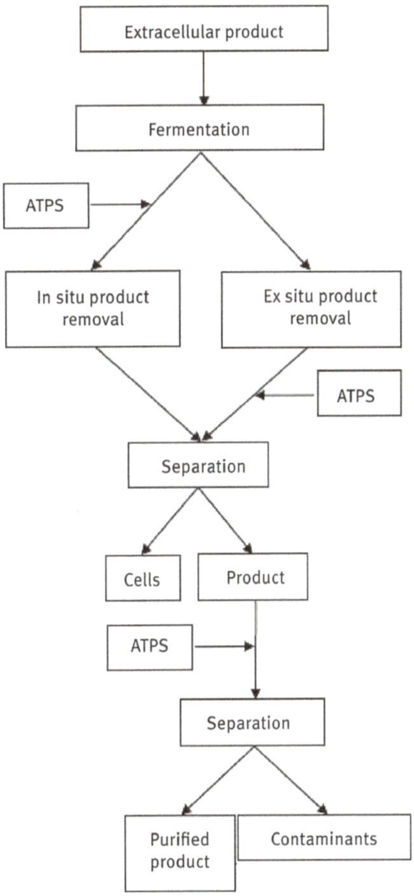

Figure 6.2: Possible application of ATPS in the process of bioproducts separation.

6.1 Polymer/polymer ATPS for separation of organic acids

Historically the investigations on the possibility to use ATPS for separation of organic acids produced by fermentation started in 1993 with the study of Dissing and Mattiasson [9] on the systems composed of a charged polymer polyethyleneimine

(PEI) and an uncharged polymer dextran (DEX), hydroxyethylcellulose (HEC) and polyethylene glycol (PEG). The authors reported that PEI is capable to form ATPS only in the presence of bi- or trivalent counter-ions (PEI was titrated with H_2SO_4 or H_3PO_4). The phase behavior of ATPS with PEI was very sensitive to pH and counter-ion. Systems at low pH had higher phase ratio (Vt/Vb) and smaller bottom (PEI-rich) phase. On the one hand, the systems with H_3PO_4 titrated PEI had even higher phase ratio and smaller bottom phase. On the other hand, lower pH led to lower polymer concentrations necessary to form ATPS but to increased viscosity and density of the bottom phase. The applicability of such systems for separation was verified in the case of bovine serum albumin (BSA). The partition experiments with BSA clearly showed that substantial changes in the partition behavior can be achieved with small changes in system composition. The authors concluded that the salt- and pH-dependencies raise the possibility of manipulating such systems, a property that might be of value in biotechnology.

In most cases, fermentative production of organic acids is product-inhibited, so simultaneous product removal is indispensable for reducing end-product inhibition and increasing process productivity. Therefore, as ATPS ensure mild and gentle conditions for the cells and cells are usually distributed in one of the phases, it is necessary to find a system where the inhibitory product is predominantly concentrated in the opposite phase. Dissing and Mattiasson [10] investigated one such system composed of 3% PEI and 7% HEC (w/w) for the cultivation of *Lactococcus lactis* and lactic acid production. The authors reported that the cultivation of *L. lactis* cells in PEI/HEC ATPS was successful with the difference that a longer lag phase was observed. Having in mind that the presence only of HEC in the fermentation broth did not affect in any way the growth of cells and lactic acid production it was obvious that the longer lag phase must be attributed to the presence of PEI. The reference cultivation without a two-phase system at higher substrate concentration (20 g/l glucose) was inhibited, whereas similar cultivation performed in ATPS showed an increase in biomass and product concentrations. The study of the distribution of cells in PEI/HEC ATPS showed that the cells avoided the bottom PEI-rich phase and about 90% of the cells were situated in the upper HEC-rich phase or on the phase boundary at pH 5.5. The obtained phase ratio was 0.9–0.95 and the distribution coefficient (K_{LA}) for lactic acid was 0.7–0.8 what means that cells and inhibitory product were predominantly situated in a different phase. Since the protonation of PEI increases at lower pH, this should increase its ability to form ion-pairs with lactate ions, but with lowering the pH lactate ions become protonated, so an optimum pH in regard of lactic acid distribution must exist. A possible restriction may be the fact that PEI adsorbs on the cells' surface and it is toxic for them. In the particular case of cultivation in PEI/HEC ATPS the cells are situated mostly in the upper HEC-reach phase what means that PEI influence was negligible. The studies of Dissing and Mattiasson unambiguously described the possibility of the use of PEI in the end-product-inhibited fermentations and other researchers have continued the investigation of other ATPS with PEI.

Kwon et al. [11] studied the potential of PEI as phase forming constituent in ATPS with either HEC or PEG. The experiments investigating phase volume ratio and lactic acid distribution in 4%PEI/1%HEC two-phase systems in the presence of phosphate buffer and fermentation medium revealed that an increase of the phosphate buffer concentration resulted in an increase in the phase volume ratio and a concomitant decrease in the partition coefficient of lactic acid. The similar trend in the change of both factors was observed when the concentration of the fermentation medium was varied. Contrary, the variations in phosphate concentration did not have a significant effect in a PEI/PEG (4%/4%) system. The changes in the concentration of phase-forming polymers affect both V_t/V_b ratio and K_{LA}. However, the system PEI/HEC was selected for further investigations as the better choice. The cultivation of *L. lactis* cells in the above mentioned ATPS was run in two modes – without and with pH control. In the fermentation of 20 g/l glucose as substrate, without pH control in the system without polymers, the lactic acid production ceased after 5–6 h, because of strong product inhibition and lowering of pH value. A similar trend was observed when the medium was supplemented only with HEC, while the fermentation in the presence only of PEI led to a long lag phase, due to the inhibition by the polycation. This inhibition influence of PEI was diminished in case of fermentation in the two-phase PEI/HEC system. Increasing the polymer's concentration from 4:1 to 5:1 and 5:1.3 led to three to four-fold increase in the lactic acid production, best in the latter one. In case of lactic acid fermentation done under pH-controlled conditions inoculation of ATPS with cells grown in normal medium resulted in a long lag phase and slower growth rate as compared to the system inoculated with cells grown in the two-phase system. Increasing the agitation speed from 50 to 150 rpm led to an increase in growth rate and a decrease in time of fermentation. Further increase of agitation speed to 300 rpm decreased growth rate and lactic acid production to the level of 50 rpm. Increasing phosphate concentration in ATPS fermentation usually led to amelioration of productivity and cells distribution in both pH-controlled and uncontrolled fermentation. The cells which favored the HEC-rich top phase in a fresh two-phase medium were partitioned to a significant extent in the bottom phase after fermentation and lactic acid was preferentially partitioned in the PEI-rich bottom phase.

Planas et al. [12] studied the cells distribution and growth and lactic acid production of three *Lactococcus* and one *Lactobacillus* strains in various ATPS composed of polyethylene glycol/dextran (PEG/DEX), polyethylene glycol/hydroxypropyl starch polymer (PEG/HPS), and a random copolymer of ethylene oxide and propylene oxide (EO-PO/HPS). The molecular weights of the used polymers were as follows: PEG – 4000 and 8000 Da, HPS – 100,000 and 200,000 Da, DEX– 40,000 and 500,000 Da and EO-PO – 4000 Da. In the partition experiments the distribution of precultured cells between the phases in different systems PEG/DEX, PEG/HPS and EO-PO/HPS was followed. The lactobacilli tended to partition predominantly in the top phase of PEG-containing systems with an exception for PEG8000/DEX40 where most of the cells were at the interface. In contrast, *lactococci* showed more irregular partitioning in these systems and what is more, the distribution was highly dependent on polymer concentrations. In the EO-PO/HPS systems, the *lactobacilli* and *lactococci* showed similar partition behavior. At low polymer concentrations, no bacteria were found in the top phase whereas increasing the polymer concentration led to the removal of the cells from the interface to both the top and the bottom phases. The EO-PO/HPS100 (5.5%/12%) system was chosen for investigation of lactic acid production with all four strains. In all cases, the amount of produced lactic acid was lower than that in the reference medium – from 2.2 to 14.5%. In view to investigating the long-term influence of the ATPS on the lactic acid production repeated extractive fermentation was carried out without pH control in the system EO-PO/HPS 100 with four-time replacing of the top phase. The lactate concentration at the end of fermentation was by more than 25% higher than in the reference fermentation. The same approach was used for studying the extractive fermentation in the system PEG/DEX [13]. Changes in the volume ratio and the composition of the phases were observed. The system was progressively enriched in DEX and depleted in PEG. Continuing the research on lactic acid production and partition in ATPS. Planas et al. [14] studied systems composed of EO-PO (with different ratio of EO to PO) and DEX. As top phase polymer $EO_{50}PO_{50}$ (4, 6 and 8%) and $EO_{30}PO_{70}$ (6.5 and 8%) were taken and DEX (Mw 500,000 Da) was the bottom phase polymer at concentrations of 4, 6, 8 and 10%. Four levels of pH – 2.0, 2.8, 3.4, and 5.0 were studied. Low- (2000 Da) and high-molecular weight (25,000 Da) PEI were used for lactic acid titration in some systems. Analysis of variance design was used to investigate the effect of pH, polymer concentration and addition of PEI to the ATPSs. In the systems without PEI the polymer concentration had no significant effect on the distribution coefficient of lactic acid (K_{lac}). The values of the distribution coefficient were significantly influenced by the pH values. In the systems with $EO_{50}PO_{50}K_{lac}$ decreased from 1.11 to 0.71 when the pH increased from 2.0 to 5.0 at low polymer concentrations. At high polymer concentrations, this decrease was more pronounced – from 1.29 to 0.46. The same trend of K_{lac} was kept when $EO_{30}PO_{70}$ was used. In case of presence of PEI in the two-phase system, the decrease in K_{lac} (systems with $EO_{30}PO_{70}$) was steeper – from 1.04 at pH 2.8 (PEI concentration 2.2%) at medium and high polymer concentrations to 0.09 at pH 6.0 (PEI concentration 7.2%) at low and medium polymer concentrations. PEI also partitioned between the phases and there was a linear relationship between K_{lac} and K_{PEI}. Titration of the produced acid with a base is a common approach to avoid the negative effects in end-product-inhibited fermentations. The authors used PEI as a titrating agent and compared the obtained results with similar fermentations where NaOH and KOH were used. In batch fermentations with 100 g/ l glucose as a substrate, in all cases the cumulative lactate concentration was above 75 g/l. The cumulative volumetric productivity showed two different patterns. When NaOH and KOH were used, the cumulative productivity increased to about 4 g/l h within 6 h, and then decreased to about 3–2 g/l h at the end of fermentation. When PEI was used for titration, the maximum productivity of about 3 g/l h was achieved within 11 h and was 2.6 g/l h at the end. The obtained results for fermentative lactic acid production in ATPS EO-PO using PEI as a titrating agent were better than those where PEI titrated with H_2SO_4 was used as phase-forming polymer together with HEC. In the further attempts to overcome constraints in the applicability of ATPSs for the recovery of lactic acid Planas et al. [15] synthesized PEG-PEI and EOPO-PEI conjugates and investigated their applicability for use in ATPS. Three different conjugates were synthesized – one EOPO-PEI with weight ratio 8:1 and two PEG-PEI with weight ratios 2:1 and 4:1. They were used for the study of lactic acid separation in ATPS with DEX and hydrolyzed maize starch (HMS). The influence of pH and the presence of phosphate and *Lactococcus* cells in the system were investigated. The partition coefficient for the cells in the systems 10% EOPO-PEI titrated with lactic acid, 8% DEX, and 2% phosphate was 0.45. In the systems with PEG-PEI/DEX lactic acid was predominantly distributed in the upper PEG-PEI conjugate-rich phase in contrast with the systems PEG/DEX where lactic acid was evenly distributed [16] or was in the bottom phase [13]. In the ATPS containing 10.0% PEG-PEI (2:1) and 8.0% DEX at pH 4.3, 4.9 and 6.4 the volume ratio was 0.6, 0.7, 0.6 and K_{lac} – 1.0, 0.99 and 0.9. The addition of 2% phosphate to the system inversed the phase volume ratio and increased the partition coefficient to values of 2.0, 2.0, 1.7 and 1.66, 1.54, 1.22, respectively. Because K_{lac} was shown to be independent of lactic acid concentration in the systems, the extractive capacity in terms of the total amount of lactic acid contained in the top phase increased with the decrease in the mass ratio between the two polymers in the conjugate. Distribution of other acids – propionic, citric and succinic also was studied in ATPS. The phenomenon of inversion of partition coefficient was observed changing the top phase forming a polymer from EOPO to EOPO-PEI. The corresponding values increased from 0.25 to 0.88 (for the system 10% EOPO–8% DEX) to 1.12–1.30 (for the system10% EOPO-PEI–8% DEX). The conjugate acted as a liquid ion-exchanger which electrostatically attracted the dominating negatively charged acid ions to the top phase. There was a certain relation between the pK_a values of the acids and the partition coefficient.

In view to develop and test biosensors for online monitoring of glucose and lactate Min et al. [17] have studied lactic acid production in an ATPS composed of 5.5% to 12% UCON 50-HB-5100 (random copolymer of ethylene oxide and propylene oxide with average Mw 4000 kD) and Reppal 200 (hydroxypropyl starch with average Mw 200 kD). Maximum lactate production rate of 0.9 g/l h was achieved, resulting in about 4 g/l lactic acid produced from 10 g/l glucose substrate.

Katzbauer et al. [16] studied the continuous fermentative lactic acid production in an ATPS formed by PEG (Mw 20,000) and DEX (Mw 229,000). The system contained 7.5% PEG and 2.5% DEX with phase volume ratio (V_t/V_b) of 6.1. The integrated fermentation-extraction system was composed by fermenter, settler in which ATPS was allowed to separate and extractor. The bottom phase was recycled from settler to the fermenter and the top phase from extractor to the fermenter after extraction of lactic acid with organic amine extractants. The *Lactobacillus casei* cells were concentrated on the interface and the partition coefficient for lactic acid was about 1. In a typical continuous fermentation with ATPS, the initial glucose concentration was 100 g/l, the dilution rate (D) $-0.09\,h^{-1}$, and recycle ratio (R) was set to 1.0 in the beginning. During the experiment the value of R was varied. Firstly, in the period from 12 h to 55 h, it was increased to 3.6 in order to enforce cell recycle and between 35 h and 50 h was decreased to 1.6. In the period between 55 h and 75 h R was kept constant at 2.5 and before stopping the experiment (103 h) it was again increased to 3.2. Maximum productivity was achieved between 50 h and 60 h. Lactic acid concentration was about 46 g/l, corresponding to a productivity of 4.2 g/l h.

In a series of works, Yankov et al. studied the systems composed of PEI (Mw 4000) and PEI (Mw 25,000). They investigated the influence of pH, type of titrating acid and amount of lactic acid in the system on the phase behavior of the system and partition of the lactic acid. In the first paper [18] a UNIQUAC model incorporating the polydispersity of PEI was found to give a good agreement between the measured and calculated compositions of the equilibrium phases. As it was pointed out above, PEI was capable to form ATPS only in case that it was titrated with bi- or trivalent counter-ions [9]. The PEI stock solutions were titrated with H_2SO_4 at pH 5.3, 7.5, or 9.2. The experimental results showed that the polymers are distributed unevenly in the two phases, with PEI concentrated in the bottom phase and PEG in the top phase. This suggested that organic acids (e. g. lactic acid) will partition preferentially in the PEI-rich phase through acid-base association. It was also observed that molar mass distribution of polydisperse PEI in both phases differed from that in the feed. The observation of Dissing and Mattiasson [9] for the strong influence of pH (degree of protonation of PEI) on the volume phase ratio was verified. In the next paper [19] the study of PEG/PEI ATPS was broadened by comparison of data for the phase behavior when PEI was titrated with a different acid and one more acidic component was added to the system. PEI is a strong polybase and its behavior markedly depends on pH. At low pH values, mutual charge repulsion leads to expansion of the polyion, while in the higher pH range the polymer contracts due to hydrogen bonding. These changes in the molecular conformation will certainly affect the interaction of the charged polymer with solvent and other constituents of the ATPS. Introducing a second acid solute (lactic acid) in the system makes it even more complex, and further complicates its phase behavior. The binodal curves were prepared at three levels of pH – 6.5, 7.5 and 9.2 with PEI titrated with H_2SO_4 or H_3PO_4 in the absence or presence of lactic acid (about 4 or 8 g/l) in the systems. The experiments showed that increasing the pH led to a contraction of the two-phase regions and that titrating PEI with a higher polyvalent acid resulted in a larger two-phase region. Since experiments showed that the lower the pH value, the wider is the immiscibility gap, it was initially expected that the addition of LA would further enhance the two-phase region of the ATPS, but in reality, the reverse was found. Finally, it was shown that the lactic acid partition coefficient (within the frame of

0.68–0.33) was very favorable (the acid was partitioned preferentially in the PEI-rich phase), which together with the fact that cells usually avoid PEI-rich phase, makes the use of PEG/PEI ATPS in the extractive lactic acid production very attractive.

A possible mechanism explaining the observed phase behavior was proposed. The presumption was that the change in PEI molecular conformation alone is not sufficient to lead to two-phase formation and hence, there is another factor, most probably related to the nature of the titrating acid. The authors suggested that the addition of polyvalent anions, as part of the titrating acid, influenced the ability of PEI molecules to form aggregates either by H-bonding or by cross-linking. Aggregation of PEI molecules would clearly enhance the incompatibility of the two polymers and would lead, in its turn, to an expansion of the two-phase region. Obviously, different titrating acids promoted to a different extent the aggregation process. In this sense, phosphoric acid had a more pronounced influence, and hence, the resulting ATPS had a larger two-phase region than when sulfuric acid was used. Returning to the influence of pH alone, it should be noted that the additional protonation of PEI (low pH values) favored aggregation and led to an increase in area of the two-phase region. These arguments were consistent with the experimental evidence. Within the framework of the mechanism discussed, the negative influence of LA on the ATPS phase formation may be attributed to destabilization of the PEI polymer aggregates, probably as a consequence of displacement of the polyvalent acid anions. As a result, the ability of the polymer to aggregate formation and cross-linking was decreased. Obviously, in the case of the H_3PO_4 titration, the influence of LA was more pronounced and led to a greater contraction of the two-phase region. Further confirmation of these arguments was provided by additional experiments in which H_2SO_4 was replaced as the titrating acid by a mixture of HCl and H_2SO_4. The latter led to a contraction of the two-phase region exactly as observed in the experiments when adding a second monovalent acid (lactic acid). The advocated mechanism was further validated with experimental data for systems where PEI was titrated with organic acids with different structure in the absence and presence of lactic acid [20]. An interesting observation was that the binodal curves in the case when PEI was titrated with tartaric acid showed no difference with that of H_2SO_4-titrated PEI. The organic acid used had a more pronounced stabilizing effect on shrinking of the two-phase region when lactic acid was added in comparison with inorganic acids. This stabilizing effect was attributed to the presence of OH groups in the structure of the organic acid used. Comparing the two-phase regions of systems prepared with H_2SO_4 and tartaric acid with two concentrations of the lactic acid smaller influence of the latter was observed in case of tartaric acid.

Ye et al. [21] have studied the extractive lactic acid production from kitchen garbage in PEG/DEX ATPS. *Lactobacillus plantarum BP04* strain was used and because of the lack of hydrolyzing enzymes, α-amylase and protease were added to the system. PEGs with molecular weight 6000, 10,000 and 20,000 and dextran with Mw 20,000 and 40,000 were used as phase-forming polymers. Different systems – 5%PEG/8%DEX, 8%PEG/8% DEX, 8%PEG/11%DEX were investigated. The growth of lactic acid bacteria was unaffected by the presence of polymers and phase volume ratio was almost constant during the process. The average rate of lactic acid production in the two-phase system was 0.2 g/l.h compared to 0.68 g/l.h in the control. Lactic acid production was independent of changes in concentration and molecular weight of PEG, while the increase in DEX molecular weight from 20,000 to 40,000 Da led to a decrease in final lactic acid concentration of about 30%. Repeated batch fermentation was realized by changing the upper phase in the 6% PEG10000/8%DEX20,000 system. After 36 h of fermentation, the system was transferred into a sterile container and allowed to split in two phases. The upper phase was replaced with fresh one supplemented with substrate and enzymes. Then phase separation was performed every 24 h for a total duration of the process of 144 h. After the first batch the concentration of produced lactic acid was about 30 g/l, decreasing to about 12 g/l in the next three batches and to about 9 g/l in the last batch. The repeated batch mode shortened the total duration of the process because the bacteria remained in the system and did not go through a lag phase.

All above-mentioned ATPS have used lactic acid as a model compound. Lactic acid fermentation is well known as a product-inhibited process and ATPSs are a prospective approach for overcoming the inhibition. Nevertheless, separation of other acids in the ATPS was also investigated.

Pereira et al. [22] studied the separation of clavulanic acid (CA, bicyclic β-lactam compound made up of a β-lactam ring and an oxazolidine ring) in different ATP systems composed of PEG and sodium polyacrylate (NaPA) with the addition of NaCl or Na_2SO_4. CA was produced by fermentation of soy oil and soy flour extract by *Streptomyces clavuligerus*. After the end of fermentation, the broth was centrifuged and the supernatant was used for partitioning experiments. For the investigation of the stability of CA different systems were evaluated. The systems were prepared at concentrations of 5 and 15% using PEGs with Mw 2000, 4000 and 6000 Da and NaPA 8000 Da. NaCl (1.05%) and Na_2SO_4 (6%) were added to each system. It was shown that the presence of polymers and salts did not considerably affect the stability of CA. On the basis of the stability tests PEG 4000 was selected and systems with PEG concentrations ranged between 4 and 21% PEG and 4 and 20% NaPA in the presence of NaCl and 2 to 10% PEG and 6 to 22% NaPA in the presence of Na_2SO_4 were prepared for partition study. High partition coefficient (above 10) was achieved in the system with 10% PEG, 20% NaPA and 6% Na_2SO_4.

Cao et al. [23] have studied the bioconversion of penicillin G to 6-aminopenicillanic acid (6APA) in PEG/dextran ATPSs, containing recombinant *Escherichia coli* cells. The cells were immobilized by crosslinking with glutaraldehyde with the goal to prevent cells breakage. Nine ATPS with PEG 6000, 10,000 and 20,000 (4 to 8.3%) and DEX T70, T500 and T2000 (3.1 to 4.8%) were investigated to determine the best conditions for bioconversion. The concentration was chosen in a way to give a phase ratio of 4:1. The amount of the cells was in proportions 10:1, 10:2 and 10:3 in respect to the total volume of the system. The system composed of 4.0% PEG 20 000/3.4% dextran T 70 with a ratio of system volume to cells' mass of 10:3 was chosen in view of the best compromise between partition coefficients of penicillin G, phenylacetic acid and 6-APA. Five consecutive runs of repeated batch bioconversions were made with initial concentration of penicillin G of 7%. At the end of each run, the composition of the top phase was analyzed, the phase was removed and replaced with a fresh one of same volume and composition. Penicillin G also was added at 7% concentration calculated according to the volume of the top phase. Conversion during repeated batch bioconversions was similar to those in stand-alone bioconversion – between 0.9 and 0.99 mol/mol. After the end of bioconversion, 6-APA was separated from the top phase by crystallization with methyl butyl ketone at pH about 4.0. The obtained 6-APA was with a purity of 96% and a yield of 83%–88%.

6.2 Polymer- salt ATPS in the separation of organic acids

Polymer-salt ATPSs are usually applied for the separation of proteins and enzymes because of the possibility of fine tuning of the system's properties by appropriate choice of salt nature (kosmotropic or chaotropic) and concentration. In reference to organic acids, polymer-salt ATPS were mostly used for *in situ* extractive separation of CA from the fermentation broth. Videira and Aires-Barros [24] studied the partition of CA in PEG- potassium phosphate ATPS. The influence of PEG molecular mass, pH, and tie-line length on the partition coefficient of CA was investigated. Depending on experimental conditions, distribution coefficients in the range of 1.5 to 114 were observed and CA was concentrated in the more hydrophobic upper PEG phase despite the fact that at the investigated pH values (7.0 and 8.0) it is in the ionic form. While for the low molecular mass PEG (400 and 1000 Da) the partition coefficient is not influenced by the Mw, a decrease of the partition coefficient for higher PEG's Mw (4000 and 6000) was observed. Increasing of the tie-line length (i. e. polymer and salt concentrations) led to a considerable increase of the partition coefficient for all tested PEGs, probably due to the salting-out effect. The pH value also influenced the partition coefficient of the potassium clavulanate – higher partition coefficients were obtained with increase in pH for low molecular mass PEGs, while no changes were observed in the case of higher molecular mass PEGs.

Silva et al. [25] have broadened the investigation on PEG/phosphates ATPS for CA separation and purification. With the aid of experimental design, the authors investigated the influence of temperature, PEG's molar mass, volume ratio of the phases, PEG and salt concentrations (i. e. tie-lines length, TLL) and pH on the yield and purification of CA. The results from 2^5 factorial design showed that increasing of the molar mass of polymer had a negative effect on CA yield and pH had a negligible effect on the yield and partition coefficient in the studied range (6.5 to 7.5). Increasing the temperature led to a decrease in the yield, while the decrease in the temperature resulted in an increase of viscosity of the systems and made them difficult to handle. The temperature was excluded from the next 2^4 central composite rotational design (CCRD) and was kept at 20 °C. On the basis of the experimental results models were obtained for partition coefficient K, yield and purification factor – first-order for K and second-order for the other two dependent variables. Statistical analysis and optimization of the mathematical models showed that the optimal values of the factors were: PEG of molar mass 400, pH 6.4, TLL 42 and volume ratio 1.3. A purification factor of 1.5 and approximately 100% yield were obtained. In a later work [26] the authors studied the application of the optimized conditions for CA purification from amino acids containing fermentation broth by PEG/phosphates ATPS and ion-exchange, once again showing the applicability of the ATPS in CA purification.

Marques et al. [27, 28] studied the influence of the agitation intensity and aeration on the separation and purification of CA in extractive fermentation using PEG/phosphate ATPS. In the first paper using fractional factorial design of the experiment, the authors showed that the concentration of the polymer and salts had little influence on the CA concentration in the top PEG-rich phase and were kept at 25% level. Finally, the best results – CA concentration of 509 mg/l, productivity of 5.3 mg/L.h, partition coefficient 8.2 and 93% yield were obtained with a system composed of 25% PEG 8000, 25% phosphates at agitation intensity 240 rpm. In the second paper, the authors investigated the influence of agitation and aeration rates on ATPS extractive fermentation of CA on the basis of the results of 2^2 full fractional design and concluded that a compromise between aeration rate and agitation intensity is required to optimize CA production and extraction by ATPS-extractive fermentation [28].

Polymer/salt ATPS are also used for separation and purification of other antibiotics which are acidic compounds by nature. For example, Mokhtarani et al. [29] have investigated the partitioning of ciprofloxacin (1-cyclopropyl-6-fluoro-4-oxo-7-(piperazin-1-yl)-quinoline-3-carboxylic acid) in PEG/Na_2SO_4 ATPS. Using orthogonal central composite design based on 2^3 full factorial experimental design, the influence of temperature, salt concentration, polymer concentration and polymer molecular weight on the partitioning of ciprofloxacin was studied. The results of the model showed that the salt concentration influenced considerably the ciprofloxacin's partition coefficient. The partition coefficient decreased on increasing the Na_2SO_4 concentration. The dependence of partition coefficient on PEG concentration passed through a minimum. Initially, the value of the partition coefficient decreased with increasing concentration and reached the minimum at about 20%. Afterward, the coefficient increased with increasing PEG's concentration. The temperature had a small influence on the partition coefficient also going through a minimum, while the ciprofloxacin concentration had an opposite trend. Increasing the initial ciprofloxacin concentration led to an increase in the partition coefficient at relatively low concentrations and decreased at high concentrations.

Wei et al. [30] studied the enzymatic synthesis of cephalexin ((7R)-3-methyl-7-(α-D-phenylglycylamino)-3-cephem-4-carboxylic acid) by penicillin G acylase (PGA) from 7-aminodeacetoxicephalosporanic acid (7-ADCA) and phenylglycine methyl ester (PGME) in various PEG/salts ATPS. Different ATPS composed of PEGs with molar mass of 400, 1000 and 2000 Da at 12 to 20% w/w and salts ($MgSO_4$, $(NH_4)_2SO_4$ and a mixture of KH_2PO_4 and K_2HPO_4 at 10 to 17.5% w/w) were used for the study of the partition of reagents. The PGA was mostly retained in the salt-rich bottom phase in the systems with $MgSO_4$, while it was concentrated in the upper phase in the systems with phosphates (partition coefficient from 30 to 65) and $(NH_4)_2SO_4$ (partition coefficient from 5.9 to 14.9). The enzyme preserved about 80% activity in the system composed of 20% (w/w) PEG 400/15% (w/w) magnesium sulfate for 35 h (at pH 6.5 and 15 °C). Changing the conditions to pH 7.0, 37 °C and 20% salt led to an additional loss of activity to about 60%. Cephalexin was partitioned in the PEG-rich upper phase (partition coefficient from 3.3 to 6.7 in the systems with $MgSO_4$ and from 1.9 to 3.2 in the systems with $(NH_4)_2SO_4$, as a decrease in the coefficient was observed with the increase in PEG's molar mass. The partition coefficients for other reagents in all systems were close to one. The cephalexin yield in ATPS was 60% compared to about 20% in an aqueous system. A series of repeated batch bioconversions was realized by removing the top phase and replacing it with a fresh one with the same composition, containing the substrates. The cephalexin yield decreased from 60% to about 40% in the second and to 30% in the third bioconversion. Nevertheless, the yield was still higher than that in the aqueous system.

Bora et al. [31] investigated the distribution of various cephalosporin antibiotics in ATPS composed by PEG 600 and Na_2SO_4, $(NH_4)_2HPO_4$ and $MgSO_4$. The influence of pH, PEG and salts concentration on the distribution coefficient has been evaluated. In general, the distribution coefficient increased with increasing pH (5.0–8.0) and PEG and salts concentration from 5 to 20%. From the obtained results in the system 20% PEG/20% Na_2SO_4 at pH 8.0 a correlation between distribution coefficient and chemical structure (in terms of hydrophobicity) of the compounds has been deduced.

Andersson et al. [32] tested the applicability of PEG/ potassium phosphate ATPS for conversion of benzylpenicillin to 6-APA by penicillin acylase from *Escherichia coli*. As a result of the optimization of system composition and reaction conditions the system composed of 8.9% (w/w) PEG 20000/7.6% (w/w) potassium phosphate was chosen for repeated batch fermentations. The enzyme was almost completely retained in the bottom salt-rich phase ($K < 0.01$) of the system, while the substrate and products were distributed mainly in the top PEG-rich phase with $K = 8.3$, 1.35 and 1.7 for benzylpenicillin, 6-APA and phenylacetic acid, respectively. In five consecutive fermentations with replacing of the top phase, the specific productivity was in the range of 0.31–1.47 μmol 6-APA mg protein^{-1} min^{-1} and the yield 0.47–0.71 mol 6-APA/mol benzylpenicillin.

Khayati [33] used the response surface methodology in order to determine the optimal conditions for propionic acid extraction in PEG/ammonium sulfate ATPS. For evaluation of the effects of PEG 4000 and ammonium sulfate concentrations, propionic acid concentration, and temperature on propionic acid extraction yield, second-order central composite design was applied. The investigated intervals were 25–35% w/w, 20–30% w/w, 0.6–1.4% w/w and 25–35 °C, respectively. In the obtained second-order polynomial equation all linear terms except PEG concentration were significant, all interaction terms were non-significant and only temperature and all quadratic terms (except ammonium sulfate) were significant. The optimization of the model led to the following optimum conditions for propionic acid extraction: PEG 4000 concentration of 30.71%, ammonium sulfate concentration of 31.77%, temperature of 37.37 °C and propionic acid concentration of 0.72%. At these conditions, an extraction of about 70% was achieved.

Wu et al. [34] used PEG/Na_2SO_4 ATPS for extraction of butyric acid from the fermentation broth. Initially, Na_2SO_4 was added to the broth with the aim to precipitate cells, sugars and nitrogen compounds. The precipitate was removed by filtration and PEG was added to form ATPS. Butyric acid, acetic acid, and butanol were concentrated in the PEG-rich top phase. After phase separation, iodine solution was added to precipitate PEG and butyric acid was separated from the filtrate by distillation. Single factor experiments were made to investigate the influence of different parameters on butyric acid extraction. PEGs with a molar mass of 1000, 2000, 4000 and 6000 Da were used in the preparation of ATPS, in which the extraction yield varied from 60 to 80% with the increase in PEG's molar mass. The concentration of phase-forming components was varied from 15 to 35% for PEG 6000 and from 8 to 13% for Na_2SO_4 with best extraction yield at 25% PEG and 9% sodium sulfate. The optimal values of the temperature and pH were determined to be 30 °C (in the range from 20 to 50 °C) and 3.0 (between 2.5 and 4.5), respectively. The volume of the fermentation broth had practically no effect on the extraction efficiency – it remained constant at the level about 90% between 5 and 35% and about 75% thereafter up to 50%. The extraction yield increased from 60 to 90% in the interval between 0.5 to 1.5 h and kept a constant value up to 4.5 h. On the basis of Box–Behnken central composite design, a ternary quadratic polynomial equation was composed and the optimum parameter values were obtained: 26% w/w PEG 6000, 9.6% w/w Na_2SO_4, 36 °C and pH 3.0. At these conditions, the maximum extraction yield of 91.74% was achieved.

Yankov et al. [35] have studied the influence of lactic acid on the formation of PEG/phosphate ATPS. The studied systems were composed of PEGs with molar mass 10,000 and 20,000 Da and K_2HPO_4, KH_2PO_4 or a mixture of both with 1:1 ratio. Binodal curves for systems without lactic acid and these containing 5.3 to 53 g/l lactic acid were compared. It was established that lactic acid acted as a one-phase promoting agent. Increasing the lactic acid concentration, the two-phase region of the ATPS shrunk for fixed PEG and phosphate concentrations and the phase volume ratio changed. Thus, some systems which were in the two-phase region at lower lactic acid concentration remained in the one-phase region at higher lactic acid concentration. The lactic acid partition coefficient was close to unity until a mass lactic acid fraction of about 4% and decreased slightly beyond it.

Li et al. [36] investigated the extractive cultivation of *L. lactis* in PEG/$MgSO_4.7H_2O$ ATPS for nisin production. Systems with PEGs 1000, 2000, 4000, 6000, 8000, 10 000, and 20 000 and different $MgSO_4$ concentrations were tested and a system with composition 11% (w/v) PEG 20 000/3.5% (w/v) $MgSO_4$. $7H_2O$ with the biggest lactic acid partition coefficient was selected. Although the cells concentration in this system was only 60% of the control, the nisin concentration increased with 33%.

PEG/salts ATPS were found to be inappropriate for low molar mass organic acids (lactic, propionic and butyric) because the distribution coefficient did not differ significantly from unity, while such systems were very promising for separation of antibiotics as they were concentrated in the salt-rich bottom phase.

6.3 ATPS alcohol-salt for separation of organic acids

The salting-out effect is a well-known phenomenon and was used long ago for fractionation of proteins and enzymes purification. In a system composed of electrolyte (inorganic salt), non-electrolyte (low molar mass water-soluble alcohol) and high molar mass organic compound (protein), the ions of the dissociated salt attract the water molecules and thus, by decreasing the number of water molecules available for water-protein interaction, cause the precipitation of the latter. In recent years, alcohol/salt systems have been increasingly used for the separation of low molar mass organic acids.

Aydoğan et al. [37] have investigated the applicability of the systems composed of ethanol or 2-propanol and $(NH_4)_2SO_4$ or K_2HPO_4 for the extraction of lactic acid from water solutions. Initially, the bimodal curves for ATPS were obtained and then the extraction yield of lactic acid in systems composed of 20% w/w inorganic salt and 28% w/w alcohol was determined. Highest yield (80%) was obtained in the system ethanol/K_2HPO_4 and by the help of 2^3 central composite design the optimum composition for best yield and distribution coefficient was determined. The model predictions gave 30.23% w/w ethanol, 18.40% w/w dipotassium hydrogen phosphate, and 80 g/l lactic acid. At these conditions extraction yield of 87% and distribution coefficient of 2.26 were obtained. A similar result was achieved with fermentation broth containing 70 g/l lactic acid – 80% in an ATPS with the same composition.

Sun et al. [38] studied ATPS consisting of organic solvents (butanol, ethanol, methanol, isopropanol, and acetone) and inorganic salts (ammonium sulfate and dipotassium hydrogen phosphate) for separation of succinic acid (SA). A synthetic solution containing 60 g/l of succinic acid was used for partition studies and the best result was obtained for the system acetone/$(NH_4)_2SO_4$. The effects of the acetone and ammonium sulfate concentrations in this system were investigated. Both the partition coefficient and the extraction yield increased with increasing of the component's concentrations. The optimal concentrations found were 20% (w/w) $(NH_4)_2SO_4$ and 30% (w/w) acetone. The obtained partition coefficient and recovery in the system were 7.88 and 89.75%, respectively. A synthetic fermentation broth containing 60.15 g/l succinic acid, 15.28 g/l lactic acid and 15.02 g/l acetic acid was used for investigation of the influence of pH on the partition coefficient and recovery in the above-mentioned system. The partition coefficients of the three acids were almost constant at pH values lower than 3.5 and the maximum recovery of succinic acid (about 90%) was obtained at pH 3.0. Finally, the system 20% $(NH_4)_2SO_4$/30% acetone was used for separation of succinic acid from real, unfiltered fermentation broth containing 51.46 g/l succinic acid, 23.82 g/l acetic acid, 11.17 g/l lactic acid, and14.57 g/l glucose. After salting-out extraction, 99.03% of the cells and 90.82% of the soluble proteins, as well as approximately 90% of the glucose, 30% of the lactic acid and 15% of the acetic acid were retained in the bottom salt-rich phase. The recovery of succinic acid was 90%. Finally, a scheme including salting-out extraction, activated carbon adsorption, vacuum distillation, and crystallization was used for succinic acid recovery from the fermentation broth. A recovery yield of 65% with a purity of 95% was obtained.

Studying the succinic acid recovery, Gu et al. [39] have screened systems composed of ethanol, methanol, 2-propanol and acetone, and K_2HPO_4, KH_2PO_4, NaH_2PO_4, K_2CO_3, $(NH_4)_2SO_4$, K_3PO_4, $MgSO_4$, Na_3PO_4. Among the systems containing 20% of salts and 20% of organic solvents, these with methanol did not form ATPS with any of the eight salts, while these containing $MgSO_4$, KH_2PO_4, and Na_3PO_4 – with any of the investigated solvents. These salts possess higher solubility in the solvents. So the ATPS composed of acetone and $(NH_4)_2SO_4$ were chosen for further study. In contrast to the finding of Sun et al. [38] the maximum succinic acid recovery in the system 20% (w/w) acetone/20% (w/w) $(NH_4)_2SO_4$; was reported at pH 2. The presence of 50 g/l acid in the ATPS slightly shifted the binodals to the lower concentrations. This behavior is opposite to that reported by Yankov et al. for the systems PEG/PEI [19] and PEG/phosphates [35]. The increase in acetone and salt concentrations led to an increase in both partition coefficient and recovery of SA. By raising the salt concentration together with increasing the partition coefficient a decrease in the value of the phase volume ratio was observed. Highest value of the partition coefficient of 4.4 was obtained in the system of 22.5% (w/w) acetone/20% (w/w) $(NH_4)_2SO_4$, while the best SA recovery (84.9%) was achieved in the ATPS of 30% (w/w) acetone/15% (w/w) $(NH_4)_2SO_4$. The influence of ATPS composition on the distribution of glucose and acetic acid was also investigated. When the salt concentration increased, concentration of glucose in the bottom salt-rich phase was observed. A similar behavior was noted when increasing the acetone concentration but in a lower degree. Regarding acetic acid distribution in the acetone/$(NH_4)_2SO_4$ ATPS, its conduct followed that of the succinic acid.

Experiments with unclarified fermentation broth revealed that the cells had little effect on the partition and recovery of SA. The cells and proteins were precipitated on the phase boundary. The concentration of acetone had a more pronounced effect on the extraction of SA, while that of $(NH_4)_2SO_4$ – on the separation of glucose and proteins. Applying a system composed of 35% (w/w) acetone and 15% (w/w) ammonium sulfate for SA extraction from a real fermentation broth without cells' removal, a recovery of SA of 94.4% was achieved with simultaneous removal of glucose and acetic acid of 93.6% and 15.5%, respectively. The cells and proteins were also removed at a ratio of 98.1% and 78.5%, respectively.

Pratiwi et al. [40] also explored the applicability of various alcohol/salt ATPS for succinic acid extraction. Ten salts – K_2HPO_4, K_3PO_4, K_2CO_3, KF, $(NH_4)_2SO_4$, $C_6H_5Na_3O_7$, Na_2CO_3, NaCl, $MgSO_4$, and NH_4NO_3 and four water-miscible alcohols – *t*-butanol, 1-propanol, 2-propanol, and ethanol were tested. Among alcohols, *t*-butanol formed ATPS with K_2HPO_4 at lower concentrations. The binodal curve with 1-propanol was very similar to that of *t*-butanol. The study of the extractability of succinic acid by ATPS with 1-propanol and different salts showed that high distribution coefficients were obtained with NaCl and $(NH_4)_2SO_4$ – 2.2 and 1.76, respectively. The lower equilibrium pH of the bottom salt-rich phase resulted in a higher extractability.

Li et al. [41] investigated the separation of enantiomeric (*R*,*S*)-mandelic acid (MA) and (*R*,*S*)-α-cyclohexylmandelic acid (α-CHMA) in alcohol/salt systems with the help of sulfonated β-cyclodextrin (Sf-β-CD) as chiral selector. Ethanol, 1-propanol, and 2-propanol, as well as $(NH_4)_2SO_4$ and KH_2PO_4 were used for ATPS formation. Three Sf-β-CD with different degrees of substitution (DS) – 3.4, 1.9 and 0.6, respectively, were used. Preliminary studies showed that Sf-β-CDs had no apparent chiral recognition ability for α-CHMA, so the experiments were continued with MA. The system composed of EtOH/$(NH_4)_2SO_4$ which possessed the maximum selectivity and Sf-β-CD with DS = 1.9 as a chiral selector was chosen. Varying components' concentrations, the system with composition 30 mass % of ethanol, 15 mass % of ammonium sulfate and 20 mmol/l of Sf-β-CD was chosen for further investigations. The experiments for temperature and pH optimization led to values of 50 °C and 2.0, respectively. An extraction time of 10 min was enough for chiral separation and MA concentration in the range of 0.5–8.0 mmol/l had a smaller influence on the enantioseparation. The optimized system composition has a potential ability for production of pure MA enantiomers at chosen reaction conditions. Separation factor of 1.69 and enantiomeric excess of 16.3% were achieved.

Chawong et al. [42] studied the system 1-butanol/$(NH_4)_2SO_4$ at 30 °C for possible extraction of lactic acid. In the absence of salt, the butanol content in water was 7.5% and that of water in the butanol was 13%. Addition of $(NH_4)_2SO_4$ led to salting-out effect and butanol content in water decreased to about 0.6% and that of water in the butanol also decreased to about 5.4%. The distribution coefficient and the degree of lactic acid extraction were dependent on lactic acid and salt concentrations and varied from 0.77 to 7.69 and from 44.01% to 89.08%, respectively, when LA concentration increased from 0.1 M to 3.0 M and $(NH_4)_2SO_4$ was changed from 0 to 5 g. Increasing the initial organic:water phase volume ratio from 1:1 to 3:1 also increased the distribution coefficient and the degree of extraction. A further improvement of the lactic acid extraction was achieved when the process was carried out in stage-wise mode rather than the batch one.

Fu et al. [43] studied the salting-out extraction of various organic acids (formic, acetic, propionic, lactic, succinic and citric) with an ATPS formed by ethanol and ammonium sulfate. The influence of different process parameters on the partition coefficient was investigated. For all acids except for citric acid, the partition coefficient increased with increasing TLL at a fixed volume ratio (1:1). This increase was attributed to the enhanced salting-out effect with increasing differences in the composition of both phases. The opposite behavior of citric acid was explained by its higher hydrophobicity. The partition coefficient was found to be almost constant when increasing the phase volume ratio. The pH value was determined as the major parameter that affected the partition of the acids in the system. In general, the partition coefficient increased with decreasing pH, except for extremely low values, while acid concentration had practically no effect on the partition coefficient. It was found that the influence of the temperature was not obvious for the majority of the acids investigated (propionic acid was again an exception). In the interval between 0 and 60 °C the extraction efficiency increased to a different extent. The changes could be explained with changes in the salt solubility and the changes in the intermolecular force of the hydrogen bonds in the system. Hydrophobicity (expressed as logP) of the acids was related to the extraction efficiency. An increase in logP led to an increase in the partition coefficient. The extraction efficiency of the investigated carboxylic acids in the EtOH/$(NH_4)_2SO_4$ system was in the order: propionic acid > acetic acid > succinic acid > lactic acid > formic acid > citric acid.

Wu et al. [44] used salting-out extraction for separation of butyric acid from aqueous solution, mixtures with acetic acid and fermentation broth. Calcium chloride had the highest partition coefficient among the investigated salts – $CaCl_2$, $MgCl_2$, NaCl and KCl. The partition coefficient decreased on increasing the temperature and increased with increasing butyric acid and salt concentrations. Using the salting-out extraction in a mixture of butyric and acetic acid with the same ratio as in the real fermentation broth – 4.1, an improvement of the ratio to about 10 was achieved by adding $CaCl_2$. In the real fermentation broth, this improvement was smaller – 5.8.

6.4 ATPS with ionic liquids and deep eutectic solvents for separation of organic acids

Unlike ordinary liquids typically composed of electroneutral molecules, the ionic liquids (IL) usually consist of a large organic ion and an organic or inorganic counter-ion. They are in liquid state at room temperature or their melting point is below 100 °C. IL's are non-flammable, chemically, electrochemically and thermo-stable, with negligible vapor pressure which makes them very useful for different separation processes. In addition to their numerous applications in separation processes in chemical, pharmaceutical and biotechnological industries, only recently ILs were used in ATPS for organic acids separation.

Yue et al. [45] compared the applicability of 1-butyl-3-methylimidazolium-hexafluorophosphate ([bmim][PF_6]), 1-octyl-3-methylimidazolium tetrafluoroborate ([omim][BF_4] and 1,2-dichloroethane, hexan-1-ol and octan-1-ol in the separation of enantiomers of mandelic acid. Different β-cyclodextrins (methyl-(ME- β-CD), hydroxyethyl-(HE- β-CD),-and hydroxypropyl-(HP- β-CD)) were tested as chiral selectors and HP- β-CD was chosen. On the basis of distribution coefficients, enantioselectivity, availability and price, [bmim][PF_6] was selected for further investigations. Studies of process parameters such as pH, temperature, mandelic acid and cyclodextrin concentrations showed that low temperature (5 °C), pH (2.5) and MA concentration (0.005 g/ml) led to the highest enantioselectivity while increasing cyclodextrin concentration increased the enantioselectivity.

Pratiwi et al. [40] investigated ATPS formed by IL and various salts for succinic acid extraction. 1-Butyl-3-methylimidazolium bromide ([Bmim]Br), 1-hexy-3-methylimidazolium bromide ([Hmim]Br) and 1-octyl-3-methylimidazolium bromide ([Omim] Br) were studied and the system [Hmim]Br/$(NH_4)_2SO_4$ was selected. The comparison with alcohols/salts ATPS showed that extractability depended not on pH but on the type of the salt used. The recovery of succinic acid from the top phase was lower in the case of IL.

Tan et al. [46] used ultrasound-assisted extraction with [Bmim]Br for separation of caffeoylquinic acids (CQAs) from a solid matrix. Then ATPS with inorganic salts were formed for enrichment of different caffeoylquinic acids. By using an experimental design a quadratic model was obtained and the optimal conditions for CQAs extraction calculated were as follows: [Bmim]Br concentration 1 mol/l, liquid-solid ratio 30 ml/g and ultrasonication time 8.39 min.

Almeida et al. [47] investigated the effect of IL as additives in PEG 300/Na_2SO_4 for extraction of gallic acid; vanillic acid and syringic acid. Five 1-butyl-3-methylimidazolium based IL were tested. The best results were achieved with IL with Cl^- as counter-ion. An increase in the partition coefficient of gallic acid from 11 to 29 was observed with 5% addition of IL in the system 23%PEG300/12% Na_2SO_4. The partition coefficients for vanillic acid and syringic acid were even higher – 46 and 50, respectively, in the same system. The extraction efficiency for all acids was about 99%.

In view of constructing a three-step purification process for penicillin separation and purification, Jang et al. [48] studied the partition behavior of penicillin in the ATPS [Bmim]Br/Na_2HPO_4. The partition coefficient was very high – it reached 1000 when the difference in IL concentrations between the two phases was about 40%. Substituting PEG with imidazole-terminal PEG (IPEG) in the PEG/ Na_2HPO_4 system led to an improvement of penicillin separation. The process comprised penicillin extraction in the IPEG-rich upper phase (step I), phase separation and addition of [Bmim]PF6 to the IPEG-rich phase, thus forming a new ATPS, separating the IPEG in the bottom IL-rich phase, leaving the penicillin in the top water phase (step II) and recovery of polymer and IL in a third ATPS formed by acidification of the phase containing IPEG and IL (step 3).

Tonova et al. [49] studied the separation of lactic acid in ATPS formed with N-octyl- and N-decylsubstituted N-methylimidazolium saccharinates ([C8/10C1im][Sac]). The authors observed that the extraction yield was independent of the side chain length of the IL, as well as from LA concentration. On the other hand, the LA partition was strongly dependent on the pH changes in the system and on the presence of kosmotropic salts. Partition coefficient $K_{LA} = 5.5$ and extraction yield EY = 81% were achieved at a pH value below pK_a and $MgSO_4$ in the LA aqueous source. When a series of consecutive extractions were performed EY = 90% was achieved only in two extraction steps (with [C8C1im][Sac]). A successful back-extraction (95%) was realized with a K_2HPO_4-containing stripping solution. The authors also proposed a possible mechanism for the extraction process, considering the secondary structure of the water-saturated long side chain of IL. On the basis of ^{1}HNMR spectra they concluded that when K_{LA} is about 1, the process could be described as physical extraction while at higher K_{LA} the transport of LA was due to the solvation rather than to binding of the ILs ions and formation of complexes LA–H_2O–IL was assumed.

Claudio et al. [50] optimized the gallic acid extraction in ionic-liquid-based ATPS. Diverse IL-based ATPS composed by seven IL and Na_2SO_4, K_2HPO_4/KH_2PO_4 and K_3PO_4 were investigated and it was shown that systems with Na_2SO_4 offered a better recovery of gallic acid in the IL-rich phase. It was shown that at low pH values the non-charged form of gallic acid was preferentially partitioned in the hydrophobic IL-rich top phase, whereas its conjugate base was preferentially partitioned in the hydrophilic salt-rich bottom phase.

Deep eutectic solvents (DES) are the result of complexation between an HBD (such as quaternary ammonium salts, etc.) and a hydrogen-bond acceptor (such as sugars, alcohols, salts, etc.). They are considered as a new class of IL and have similar physicochemical properties as ILs, such as negligible volatility, high dissolving capacity, high thermal and electrochemical stabilities, and nonflammability. Moreover, DES have additional benefits like lower toxicity, lower cost, easier preparation and storage, and better biodegradability, which make them very attractive for different bioseparation processes. The first use of ATPS with DES was realized by Zheng et al. [51] for the separation of proteins.

Wang et al. [52] conducted the extraction of chlorogenic acid (CGA) from blueberry leaves by DES. Eight different DES were synthesized and tested for their extraction ability. Choline chloride-1,3-butanediol was selected because it showed higher extraction capability. Response surface methodology (RSM) was used for optimizing the values of influencing parameters – temperature, liquid/solid ratio, and extraction time. The optimized conditions were as follows: temperature 59.03 °C, liquid/solid ratio 17.01 mL/g, extraction time 24.12 min. The obtained extraction yield of CGA was 46.88 mg/g. For further enrichment of CGA an ATPS was formed by adding K_2HPO_4 to the DES. Optimum conditions for CGA enrichment were found as follows: choline chloride 1,3-butanediol concentration 26.72% (w/w), K_2HPO_4 concentration 35.11% (w/w), pH 3, temperature 35 °C and water content in DES 20%. At these conditions, the achieved recovery of CGA by DES-ATPS was 96.18%, and the purity – 74.5%.

Farias et al. [53] evaluated liquid-liquid equilibria of ATPSs composed of DES/K_2HPO_4. Various choline chloride/sugars DES were prepared at different proportions. ATPS were applied to investigate the partition of gallic acid (GA). In all studied systems, the pH was in the range 10.3–11 and there was no difference between the pH values of the top and bottom phases. Gallic acid was preferentially partitioned in the choline chloride-rich top phase. The higher values of the partition coefficient resulted in higher extraction efficiency. Differences in the partition coefficient of gallic acid between the hydrogen bond donors (HBDs) of DES were observed. The partition coefficient increased in the order: fructose < glucose < saccharose, i. e. opposite to the hydrophobicity of the sugars.

In a newer paper [54] the researchers studied the influence of pH on the phase equilibrium and DES' components partition in ATPS composed of potassium citrate ($K_3C_6H_5O_7$)/citrate buffer (citric acid ($C_6H_8O_7$) + potassium citrate and DES (tetrabutylammonium chloride ([N4444]Cl) + ethanol or n-propanol). Binodal curves of different ATPS at different pH values – $K_3C_6H_5O_7$ (for pH 9) or $K_3C_6H_5O_7/C_6H_8O_7$ buffer (for pH 7 and 5) and aqueous solutions of [N4444]Cl and alcohols mixed at different molar ratios (2:1, 1:1, and 1:2) were determined. Increasing the pH value led to the enlargement of the two-phase region. The partition coefficients of gallic acid, caffeine, and L-tryptophan were also determined at fixed TTL in a system consisting of 25 wt % of citrate salt + 30 wt % of DES. For gallic acid, the partition coefficient tends to increase with increasing pH. The changes of the partition coefficient were explained with changes in the ratio of mono- and divalent species at different pH values and their different partition into the top [N4444]Cl-rich phase. This was in agreement with findings of Claudio et al. [50] for GA partition in IL-based ATPS. Panas et al. [55] used ATPS formed by PEG or polyethylene glycol methyl ether and cholinium chloride (ChCl) for separation of CA. Firstly, the stability study of CA in ChCl aqueous solution ((2.2–4.3 M) showed that CA retained about 90% of its activity for more than 3 h. Although CA was almost evenly distributed in both phases, the partition of CA could be influenced by proper adjustment of ATPS composition. Further ChCl was used as an adjuvant to ATPS PEG600/NaPA-8000 and it was observed that CA was preferentially distributed in the PEG-rich phase with an extraction efficiency of over 85%. The same system was used for CA separation from the real fermentation broth and high purification from contaminant proteins was observed.

One of the main problems in using DES in ATPS is the weakening of the interaction between the HBD and the hydrogen bond acceptor (HBA) in the presence of water, and complete disruption of DES complexes can occur at high water content [56]. As a result, a nonstoichiometric partition of HBA and HBD between the phases of the ATPS is observed. The same phenomenon was observed by Passos et al. [57] when studying the cholinium chloride/urea system for separation of textile dyes. In an attempt to overcome this problem Farias et al. [58] studied the partition of gallic acid and vanillic acid among other biomolecules in systems composed of polypropylene glycol (PPG) and DES (cholinium chloride and glucose or urea). It was shown that the initial molar ratio (HBA:HBD) can be maintained in ATPS. Depending on the nature and concentration of the other phase forming components, and the nature of HBA and HBD used, it was possible to manipulate the HBA:HBD molar ratio in the ATPS phases. In case that both HBD and HBA were highly hydrophilic and were poorly soluble in the polymer-rich phase the DES initial molar ratio can be maintained in the DES-polymer-based ATPS. It was observed that HBD concentration and TLL also influenced the partition of the biomolecules. All these factors can be used to manipulate and fine-tune the partition of the biomolecules in such systems, as the hydrophobicity difference between the phases was the main factor governing the partition.

6.5 Surfactant-based ATPS for organic acids separation

When certain classes of surfactants aqueous solutions are heated above the so-called cloud point temperature (CPT), an ATPS is formed with two aqueous, immiscible phases – surfactant-rich and surfactant-poor. The phase separation can occur also by varying other conditions such as pH, and ionic strength. This is also true when the surfactant concentration is higher than the critical micellar concentration (CMC). Then the system separates to micellar-rich and micellar-poor phases. The amphiphilic nature of surfactants together with control and optimization of the partitioning behavior by tuning the micellar characteristics helps to adjust the separation of a target biomolecule depending on its hydrophobicity. In the work of de Andrade et al. [59], the possibility of CA extraction in a two-phase micellar system was described. The investigated surfactants were *n*-decyltetraethylene oxide ($C_{10}E_4$) and dodecyldimethylamine oxide (DDAO). CA partitioned evenly between the two phases of the DDAO micellar system, the $C_{10}E_4$ micellar system, and the mixed DDAO – $C_{10}E_4$ micellar system, with recovery not exceeding 35% for DDAO and 71% for the mixed system. Nevertheless, the application of $C_{10}E_4$ for CA separation from real *S. clavuligerus* fermentation broth led to 52% recovery and 70% removal of protein contaminants when a denaturation step was applied before extraction.

Santos et al. [60] investigated a mixed micellar system composed of a nonionic surfactant (Triton X-114) and an anionic one (AOT) for separation of CA. CA showed an excellent stability in surfactant solutions and partitioned preferentially in the top micelle-poor phase. Applying 2^4 – full factorial design the extraction parameters – temperature, surfactants, and NaCl concentrations were optimized. The regression analysis of the constructed quadratic model revealed that the partition coefficient was mainly influenced by the NaCl concentration and the interaction between temperature and Triton X-114 concentration while CA yield was mainly influenced by temperature, Triton X-114 concentration and their interaction. The system composed of 0.022% (w/w) AOT and 1% (w/w) Triton X-114 at 28 °C without NaCl was selected as the optimal one for the separation of CA from the fermentation broth.

In the next paper of the same group, Haga et al. [61] continued the study of CA extraction by means of micellar two-phase systems. Systems containing only the nonionic surfactant, $C_{10}E_4$, as well as mixed systems with cationic CTAB or anionic AOT surfactants were used. The CA partitioning in the C10E4 /buffer micellar system at pH 6.5 decreased from 0.87 to 0.67 with increasing the temperature from 21 to 28 °C. The addition of CTAB (6 to 24%) and the increase in the temperature to 64.8 °C increased the partition coefficient to about 1.4. It was anticipated that electrostatic interactions would influence the partitioning of CA in the micelle-rich phase. Contrary, the addition of AOT led to the lowest partition coefficient values. The percentage of CA recovery by the AOT/ $C_{10}E_4$ micellar system, was 43.0 ± 11.9%, whereas for the CTAB/ $C_{10}E_4$ micellar system it was 21.5 ± 6.8%. The authors concluded that further investigations were necessary to improve system conditions in order to achieve better partition coefficients and mass balances.

Searching for an efficient and cost-effective downstream process for CA separation, Silva et al. [62] investigated Triton X-114 and Triton X-100 two-phase micellar systems with or without the addition of Dextran sulfate (Dx-S). The CA stability check experiments were conducted at a temperature of 5, 20, 35 and 45 °C at pH 6.5 in systems with Dx-S from 1 to 15%. At 5 °C, CA concentration was nearly 100% after 24 h; on increasing the temperature to 20 and 35 °C, CA concentration was still above 90% after 6 h with significant loss of activity thereafter. Since Dx-S is an anionic polymer, the authors expected its addition to promote electrostatic repulsion of CA to the micellar phase of the system. Varying the Dx-S concentration in systems with Triton X-114 and Triton X-100 a partition coefficient of 1.83 was obtained in the system 14% (w/w) Dx-S/9% (w/w) of Triton X-100 at 42 °C. The experiments showed that CA could be recovered in the micelle-rich phase of a Dx-S/Triton X system as a first purification step, while a great part of the bioprocess contaminants rested in the micelle-poor phase (Dx-S rich).

Tannic acid (TA) was also used as a subject of cloud-point extraction. Ghouas et al. [63] investigated the TA removal by two nonionic surfactants – Lutensol ON 30 and Triton X-114. It was shown that by addition of an ionic substance to the nonionic surfactant solution the CPT could be manipulated. Thus, the addition of TA or CTAB increased CPT whereas Na_2SO_4 decreased it. Using four parameters (TA extraction, TA and surfactant concentrations in the dilute phase and volume fraction of the surfactant-rich phase) of the central composite design the values of the parameters of a quadratic model were determined. At temperatures of 40–45 °C and very low volume fraction of surfactant-rich phase (0.03) around 95% (for TX-114) and 87% (for Lutensol ON 30) extraction of TA was achieved. Addition of Na_2SO_4 and CTAB increased the TA extraction. Surfactant regeneration could not be realized by changing pH because the pH change showed no effect on its partition. Hydrophobic organic solvents (petroleum ether and cyclohexane) and a more hydrophilic organic solvent (diethyl ether) were used for TA back-extraction and surfactant regeneration and diethyl ether was selected.

Yao and Yang [64] developed a novel method for acetic acid (AA) extraction from dilute solutions. The new approach consisted in complex formation between AA and a complexing agent, complex solubilization in a nonionic surfactant and micelles concentration by heating over CPT. Tributyl phosphate (TBP) and tri-n-octyl/decylamine (TOA) as complexing agents and Triton X-114, Tergitol TMN-6 and PEG/PPG-18/18 Dimethicone (DC 190) as surfactants were used. Dilute aqueous solution of acetic acid (0.1 M), complexing agent and surfactant were stirred for 1 h at 60 °C, then incubated for 30 min at the same temperature and finally separated by centrifugation. In the systems with Tergitol TMN-6, no phase separation was observed while when TOA was employed as a complexing agent very low recovery of AA (<5%) was found. The distribution coefficient was close to 1 and the recovery from 62 to 69%. Further process parameters optimization was made for the system DC 190/TBP/AA. The optimal values were as follows: surfactant (DC-190) concentration – 50 wt%, complexing agent (TBP) concentration – 2 wt%, temperature for stirring – 60 °C, incubation temperature – 55 °C wt% and AA concentration – 0.1 M. At these conditions, a relatively high recovery of 71.4% and a distribution coefficient of 1.4 were achieved.

Simental-Martínez et al. [65] compared the partition behavior of gallic acid, chlorogenic acid and three other small compounds in different ATPS – polymer/polymer, polymer/salt, IL/salt, and alcohol/salt. Twenty-nine different systems were investigated. In PEG-phosphate ATPSs, the compounds preferably partitioned in the top polymer-rich phase. The partition coefficient increased with increasing molecular weight of PEG and TTL what was related to the comparative hydrophobicity of both phases. Similar behavior was also observed in the system PEG/DEX but the partition coefficients were lower when compared to the polymer/salt system. Two ionic liquids [BMIM][BF_4] and [EMIM][Ac] were used for ATPS formation with sodium and potassium phosphate, respectively. It was shown that as the hydrophobicity increased, the preference of the two acids towards the IL-rich phase also increased. It was supposed that the preference to the top phase was driven in a higher degree by hydrophobic interactions, but the lower partition coefficients of gallic acid (the smaller and more hydrophobic compound) suggested that other interactions might be taking place. In the ethanol-potassium phosphate ATPS, all compounds showed a preference for the top (ethanol-rich) phase. For chlorogenic acid, there was a relation between TLL and partition in the top phase, while gallic acid was partitioned evenly in both phases, independently of the TTL. Comparing the recovery of both acids in all studied systems it was concluded that for the chlorogenic acid higher recovery was achieved in the polymer/salt system with high-molecular-weight PEG, whereas for gallic acid the best system was[EMIM][Ac].

6.6 Conclusions

In this review, an attempt was made to summarize the achievements of using ATPS in downstream processes of low molecular mass organic acids. It is evident that ATPSs are a powerful tool for the separation and purification of organic acids as an alternative to the traditional techniques, because of their biocompatibility, selectivity, easy scale-up and possibility of use in continuous mode. By careful selection of system's type and constituents it is possible to manipulate the partition of the acids. There are some drawbacks that prevent for now the application of ATPS on a large scale. Still, the mechanism governing the two-phase formation and partition of organic acids in different types of ATPS is poorly understood. ATPS should generate large amounts of polymers and salts that could rise up environmental problems if they are not effectively recycled at low cost. It is expected novel polymers, ionic liquids, DES, as well as novel types of ATPS to be discovered thus helping to overcome some of the mentioned disadvantages. This together with a better understanding of the partition mechanism will make ATPS an indispensable part of bioproduct recovery processes.

References

[1] Beijerinck MW, Zentbl. Bakteriol. 1896;2:697–99.
[2] Albertsson PA. Partition of cell particles and macromolecules. New York: Wiley, 1986.
[3] Hatti-Kaul R. Aqueous two-phase systems a general overview. Mol Biotechnol. 2001;19:269–77.
[4] Rito-Palomares M. Practical application of aqueous two-phase partition to process development for the recovery of biological products. J Chromatogr B. 2004;807:3–11.
[5] Aguilar O, Rito-Palomares M. Aqueous two-phase systems strategies for the recovery and characterization of biological products from plants. J Sci Food Agric. 2010;90:1385–92.
[6] Raja S, Murty VR, Thivaharan V, Rajasekar V, Ramesh V. Aqueous two phase systems for the recovery of biomolecules – a review. Sci Technol. 2011;1:7–16.
[7] Goja AM, Yang H, Cui M, Li C. Aqueous two-phase extraction advances for bioseparation. J Bioproces Biotechniq. 2013;4, Article ID 1000140:1–8.
[8] Iqbal M, Tao Y, Xie S, Zhu Y, Chen D, Wang X, et al. Aqueous two-phase system (ATPS): an overview and advances in its applications. Biol Proced Online. 2016;18:1–18. DOI: 10.1186/s12575-016-0048-8.
[9] Dissing U, Mattiasson B. Poly(ethyleneimine) as a phase-forming polymer in aqueous two-phase systems. Biotechnol Appl Biochem. 1993;17:15–21.
[10] Dissing U, Mattiasson B. Cultivation of *Lactococcus luctis* in a polyelectrolyte neutral polymer aqueous two-phase system. Biotechnol Lett. 1994;16:333–8.

[11] Kwon YJ, Kaul R, Mattiasson B. Extractive lactic acid fermentation in poly(ethyleneimine)-based aqueous two-phase system. Biotechnol Bioeng. 1996;50:280–90.
[12] Planas J, Rådström P, Tjerneld F, Hahn-Hägerdal B. Enhanced production of lactic acid through the use of a novel aqueous two-phase system as an extractive fermentation system. Appl Microbiol Biotechnol. 1996;45:737–43.
[13] Planas J, Lefebvre D, Tjerneld F, Hahn-Hägerdal B. Analysis of phase composition in aqueous two-phase systems using a two-column chromatographic method: application to lactic acid production by extractive fermentation. Biotechnol Bioeng. 1997;54:303–11.
[14] Planas J, Varelas V, Tjerneld F, Hahn-Hägerdal B. Amine-based aqueous polymers for the simultaneous titration and extraction of lactic acid in aqueous two-phase systems. J Chromatogr B. 1998;711:265–75.
[15] Planas J, Kozlowski A, Milton Harris J, Tjerneld F, Hahn-Hägerdal B. Novel polymer–polymer conjugates for recovery of lactic acid by aqueous two-phase extraction. Biotechnol Bioeng. 1999;66:211–8.
[16] Katzbauer B, Cesi V, Narodoslawsky M, Moser A. Extractive lactic acid fermentation using aqueous two-phase systems. Chem Biochem Eng Q. 1995;9:79–87.
[17] Min RW, Rajendran V, Larsson N, Gorton L, Planas J, Hahn-HaÈgerdal B. Simultaneous monitoring of glucose and L-lactic acid during a fermentation process in an aqueous two-phase system by on-line FIA with microdialysis sampling and dual biosensor detection. Anal Chim Acta. 1998;366:127–35.
[18] Yankov DS, Stateva RP, Martin Trusler JP, St Cholakov G. Liquid-liquid equilibria in aqueous two-phase systems of poly(ethylene glycol) and poly(ethyleneimine): experimental measurements and correlation. J Chem Eng Data. 2006;51:1056–61.
[19] Yankov DS, Martin Trusler JP, Stateva RP, St Cholakov G. Influence of pH and acid solutes on the phase behaviour of aqueous solutions containing poly(ethylene glycol) and poly (ethyleneimine). Biochem Eng J. 2009;48:104–10.
[20] Yankov DS, Beschkov VN, Stateva RP. Influence of acid solutes on the phase behaviour of aqueous two-phase systems, containing poly(ethylene glycol) and poly(ethylene imine). Bulg Chem Commun. 2010;42:327–34.
[21] Ye Z, Cai W, Lu M. Lactic acid extraction from kitchen wastes fermentation using aqueous two-phase systems. Chin J Environ Eng. 2016;6:3299–304.
[22] Pereira JF, Santos VC, Johansson H-O, Teixeira JA, Pessoa Jr A. A stable liquid–liquid extraction system for clavulanic acid using polymer-based aqueous two-phase systems. Sep Purif Technol. 2012;98:441–50.
[23] Cao X-J, Wu X-Y, Fonseca LJ, Cabral JM, Marcos JC Production of 6-Aminopenicillanic acid in aqueous two-phase systems by recombinant Escherichia coli with intracellular penicillin acylase. Biotechnol Lett. 2004;26:97–101.
[24] Videira M, Aires-Barros MR. Liquid-liquid extraction of clavulanic acid two-phase system of polyethylene glycol phosphate. J Chromatogr A. 1994;668:237–40.
[25] Silva CS, Bovarotti E, Rodrigues MI, Hokka CO, Barboza M. Evaluation of the effects of the parameters involved in the purification of clavulanic acid from fermentation broth by aqueous two-phase systems. Bioprocess Biosystem Eng. 2009;32:625–32.
[26] Silva CS, Cuel MF, Barreto VO, Kwong WH, Hokka CO, Barboza M. Separation of clavulanic acid from fermented broth of amino acids by an aqueous two-phase system and ion-exchange adsorption. N Biotechnol. 2012;29:428–31.
[27] Marques DA, Pessoa-Junior A, Lima-Filho JL, Converti A, Perego P, Porto AL. Extractive fermentation of clavulanic acid by streptomyces DAUFPE 3060 using aqueous two-phase system. Biotechnol Prog. 2011;27:95–103.

[28] Marques DA, Santos-Ebinuma VC, Pessoa-Junior A, Porto AL, Torres BR, Converti A. Effect of aeration and agitation on extractive fermentation of clavulanic acid by using aqueous two-phase system. Biotechnol Prog. 2016;32:1444–52.

[29] Mokhtarani B, Karimzadeh R, Amini MH, Manesh SD. Partitioning of Ciprofloxacin in aqueous two-phase system of poly(ethylene glycol) and sodium sulphate. Biochem Eng J. 2008;38:241–7.

[30] Wei D-Z, Zhu J-H, Cao X-J. Enzymatic synthesis of cephalexin in aqueous two-phase systems. Biochem Eng J. 2002;11:95–9.

[31] Bora MM, Borthakur S, Rao PC, Dutta NN. Aqueous two-phase partitioning of cephalosporin antibiotics: effect of solute chemical nature. Sep Purif Technol. 2005;45:153–6.

[32] Andersson E, Mattiasson B, Hahn-Hägerdal B. Enzymatic conversion in aqueous two-phase systems: deacylation of benzylpenicillin to 6-aminopenicillanic acid with penicillin acylase. Enzyme Microb Technol. 1984;6:301–6.

[33] Khayati G. Optimization of propionic acid extraction by aqueous two-phase system using response surface methodology. Chem Eng Commun. 2013;200:667–77.

[34] Wu X, Li G, Yang H, Zhou H. Study on extraction and separation of butyric acid from clostridium tyrobutyricum fermentation broth in PEG/Na2SO4 aqueous two-phase system. Fluid Phase Equilib. 2015;403:36–42.

[35] Yankov DS, Trusler JP, Yordanov BY, Stateva RP. Influence of lactic acid on the formation of aqueous two-phase systems containing poly(ethylene glycol) and phosphates. J Chem Eng Data. 2008;53:1309–15.

[36] Li C, Ouyang F, Bai J. Extractive cultivation of *Lactococcus lactis* using a polyethylene glycol/$MgSO_4.7H_2O$ aqueous two-phase system to produce nisin. Biotechnol Lett. 2000;22:843–7.

[37] Aydoğan Ö, Bayraktar E, Mehmetoğlu Ü. Aqueous two-phase extraction of lactic acid: optimization by response surface methodology. Sep Sci Technol. 2011;46:1164–71.

[38] Sun Y, Yan L, Fu H, Xiu Z. Salting-out extraction and crystallization of succinic acid fromfermentation broths. Process Biochem. 2014;49:506–11.

[39] Gu BH, Zheng P, Yan Q, Liu W. Aqueous two-phase system: an alternative process for recovery of succinic acid from fermentation broth. Sep Purif Technol. 2014;138:47–54.

[40] Pratiwi AI, Yokouchi T, Matsumoto M, Kondo K. Extraction of succinic acid by aqueous two-phase system using alcohols/salts and ionic liquids/salts. Sep Purif Technol. 2015;155:127–32.

[41] Li F-F, Tan Z-J, Guo Z-F. Enantioseparation of mandelic acid and α-cyclohexylmandelic acid using an alcohol/salt-based aqueous two-phase system. Chem Pap. 2014;68:1539–45.

[42] Chawong K, Rayabsri C, Rattanaphanee P. Extraction of lactic acid in mixed solvent electrolyte system containing water, 1-butanol and ammonium sulfate. Int J Chem React Eng. 2015;13:183–8.

[43] Fu H, Sun Y, Teng H, Zhang D, Xiu Z. Salting-out extraction of carboxylic acids. Sep Purif Technol. 2015;139:36–42.

[44] Wu D, Chen H, Jiang L, Cai J, Xu Z, Cen P. Efficient separation of butyric acid by an aqueous two-phase system with calcium chloride. Chin J Chem Eng. 2010;18:533–7.

[45] Yue Y, Jiang X-Y, Yu J-G, Tang K-W. Enantioseparation of mandelic acid enantiomers in ionic liquid aqueous two-phase extraction systems. Chem Pap. 2014;68:465–71.

[46] Tan T, Lai C-J, Yang HO, He M-Z, Feng Y. Ionic liquid-based ultrasound-assisted extraction and aqueous two-phase system for analysis of caffeoylquinic acids from Flos Lonicerae Japonicae. J Pharm Biomed Anal. 2016;120:134–41.

[47] Almeida MR, Passos H, Pereira MM, Lima AS, Coutinho JA, Freire MG. Ionic liquids as additives to enhance the extraction of antioxidants in aqueous two-phase systems. Sep Purif Technol. 2014;128:1–10.

[48] Jiang Y, Xia H, Yu J, Guo C, Liu H. Hydrophobic ionic liquids-assisted polymer recovery during penicillin extraction in aqueous two-phase system. Chem Eng J. 2009;147:22–6.
[49] Tonova K, Svinyarov I, Bogdanov MG. Hydrophobic 3-alkyl-1-methylimidazolium saccharinates as extractants for L-lactic acid recovery. Sep Purif Technol. 2014;125:239–46.
[50] Cláudio AF, Ferreira AM, Freire CS, Silvestre AJ, Freire MG, Coutinho JA. Optimization of the gallic acid extraction using ionic-liquid-based aqueous two-phase systems. Sep Purif Technol. 2012;97:142–9.
[51] Zeng Q, Wang Y, Huang Y, Ding X, Chen J, Xu K. Deep eutectic solvents as novel extraction media for protein partitioning. Analyst. 2014;139:2565–73.
[52] Wang T, Xua W-J, Wang S-X, Koua P, Wang P, Wang X-Q, et al. Integrated and sustainable separation of chlorogenic acid from blueberry leaves by deep eutectic solvents coupled with aqueous two-phase system. Food Bioprod Process. 2017;105:205–14.
[53] Farias FO, Sosa FH, Igarashi-Mafra L, Coutinho JA, Mafra MR. Study of the pseudo-ternary aqueous two-phase systems of deep eutectic solvent (choline chloride: sugars)+ K_2HPO_4 + water. Fluid Phase Equilib. 2017;448:143–51.
[54] Farias FO, Passos H, Coutinho JA, Mafra MR. pH effect on the formation of deep-eutectic-solvent-based aqueous two-phase systems. Ind Eng Chem Res. 2018;57:16917–24.
[55] Panas P, Lopes C, Cerri MO, Ventura SP, Santos-Ebinuma VC, Pereira JF. Purification of clavulanic acid produced by *Streptomyces clavuligerus* via submerged fermentation using polyethylene glycol/cholinium chloride aqueous two-phase systems. Fluid Phase Equilib. 2017;450:42–50.
[56] Dai Y, Witkamp G-J, Verpoorte R, Choi YH. Tailoring properties of natural deep eutectic solvents with water to facilitate their applications. Food Chem. 2015;187:14–9.
[57] Passos H, Tavares DJ, Ferreira AM, Freire MG, Coutinho JA. Are aqueous biphasic systems composed of deep eutectic solvents ternary or quaternary systems. ACS Sustainable Chem Eng. 2016;4:2881–6.
[58] Farias FO, Passos H, Lima AS, Mafra MR, Coutinho JA. Is it possible to create ternary-like aqueous biphasic systems with deep eutectic solvents. ACS Sustainable Chem Eng. 2017;5:9402–11.
[59] de Andrade EG, Silva MS, Haga RB, Santos VC, Pessoa Jr A, Rangel-Yagui CO. Extraction of clavulanic acid using aqueous two-phase micellar system. Biotechnol Appl Biochem. 2011;58:103–8.
[60] Santos VC, Hasmann FA, Converti A, Pessoa Jr A. Liquid–liquid extraction by mixed micellar systems: A new approach for clavulanic acid recovery from fermented broth. Biochem Eng J. 2011;56:75–83.
[61] Haga RB, Santos-Ebinuma VC, Silva MS, Pessoa Jr A, Rangel-Yagui CO. Clavulanic acid partitioning in charged aqueous two-phase micellar systems. Sep Purif Technol. 2013;103:273–8.
[62] Silva MS, Santos-Ebinuma VC, Lopes AM, Rangel-Yagui CO. Dextran sulfate/Triton X two-phase micellar systems as an alternative first purification step for clavulanic acid. Fluid Phase Equilib. 2015;399:80–6.
[63] Ghouas H, Haddou B, Kameche M, Canselier JP, Gourdon C. Removal of tannic acid from aqueous solution by cloud point extraction and investigation of surfactant regeneration by microemulsion extraction. J Surfact Deterg. 2016;19:57–66.
[64] Yao B, Yang L. Cloud point extraction of acetic acid from aqueous solution. Sep Sci Technol. 2009;44:476–90.
[65] Simental-Martínez J, Montalvo-Hernández B, Rito-Palomares M, Benavides J. Application of aqueous two-phase systems for the recovery of bioactive low-molecular weight compounds. Sep Sci Technol. 2014;49:1872–82.

Konstantza Tonova

7 Ionic liquid-assisted biphasic systems for downstream processing of fermentative enzymes and organic acids

Abstract: Room-temperature ionic liquids (ILs) represent molten salts entirely consisting of ions, usually a charge-stabilized organic cation and an inorganic or organic anion. ILs are liquids at ambient temperature but possess characteristics unusual for the common liquid solvents, such as negligible vapor pressure, high thermal stability and most over the ability to mix and match libraries of cations and anions in order to acquire desirable physical and chemical properties [1]. The opportunity to obtain tunable density, viscosity, polarity and miscibility with common molecular liquids gave rise to a variety of applications of the ILs [2] as environmentally benign solvents, extractants or auxiliaries. In particular, numbers of innovations in the methods for recovery and purification of biologically derived compounds involve ILs used solo or partnered with other liquids in biphasic systems [3–5]. It should be noted that the ILs are not intrinsically greener than the traditional solvents, given that their production is usually more resource-demanding, but the inherent potential for recycling and reuse, and for prevention of chemical accidents gives the ILs advantages ahead.

The present chapter provides a state-of-the-art overview on the basic applications of the ILs in biphasic systems aimed at downstream processing of valuable fermentative products, enzymes and organic acids. Main industrially important enzymes, lipases and carbohydrases, are considered and a description of the IL-assisted aqueous biphasic systems (ABS) and the results obtained in view of enzyme yield and purity is made. ILs serve different functions in the ABS, main phase-segregating constituents (mostly in the IL/salt ABS) or adjuvants to the polymer/salt ABS. Enzyme isolation from the contaminant proteins present in the feedstock can be carried out either in the IL-rich or in the salt-rich phase of the ABS and for the reader's convenience the two options are described separately. Discussion on the factors and parameters affecting the enzyme partitioning in the ABS with ILs guides the reader through the ways by which the interactions between the IL and the enzyme can be manipulated in favor of the enzyme purification through the choice of the ABS composition (IL, salt, pH) and the role of the water content and the IL-rich phase structure.

This article has previously been published in the journal Physical Sciences Reviews. Please cite as: Tonova, K. Ionic liquid-assisted biphasic systems for downstream processing of fermentative enzymes and organic acids *Physical Sciences Reviews* [Online] 2021, 1. DOI: 10.1515/psr-2018-0068

https://doi.org/10.1515/9783110574111-007

The second part of the chapter is dedicated to the recovery of fermentative organic acids. Mostly hydrophobic ILs have been engaged in the studies and the biphasic systems thereof are summarized. The systems are evaluated by the extraction efficiency and partition coefficient obtained. Factors and parameters affecting the extraction of organic acids by ILs are highlighted in a way to unravel the extraction mechanism. The choice of IL and pH determines the reactive mechanism and the ion exchange, while the water content and the IL phase structure play roles in physical extraction. Procedures undertaken to enhance the efficiency and to intensify the process of extraction are also looked over.

Finally, the experimental holes that need fill up in the future studies are marked. According to the author's opinion an intense research with hydrophobic ILs is suggested as these ILs have been proved milder to the biological structures (both the microbial producer and the enzyme product), more effective in the organic acid recovery and suitable to perform "in situ" extraction. Extractive fermentation entails validation of ecological and toxicological characteristics of the ILs. The protocols for re-extraction of fermentative products separated by IL-assisted biphasic systems should be clearly settled along with the methods for ILs recycling and reuse. Novel more flexible approaches to process intensification can be implemented in order to adopt the separation by biphasic systems for use in industry.

Keywords: biphasic system, enzyme, ionic liquid, mechanism, organic acid, purification, recovery

7.1 Enzyme recovery and purification by ABS with ILs

7.1.1 Overview

7.1.1.1 Lipolityc enzymes based on the example of lipase

Lipase is a versatile catalyst that operates at the "water/oil" interface having substrates of different polarities in each of the reactions, triacylglycerols and water in hydrolysis, and fatty acid and hydrophilic alcohol in synthesis (esterification). The effective reaction systems for heterogeneous biocatalysis should provide high interfacial contact area between the enzyme and the substrate(s), fast exchange processes and prevention of enzyme aggregation. Nano-size structured fluids have been associated with the idea of a micro-reactor where the problem of solubility of the substances with different polarity and their contact with the enzyme catalyst could be solved on a molecular scale. Such liquid media are represented by reversed micellar solvents formed by ionic or non-ionic surfactants in nonpolar organic solvents [6, 7] and surface-active ILs with long alkyl side chain in ABS (aqueous biphasic systems). Here, the latter systems are considered as a tool for gaining lipase isolation and purification from fermentation media.

7.1.1.1.1 Lipase recovery in the ionic liquid (IL)-rich phase of the ABS

Deive and co-workers first studied the IL-assisted extraction of lipase and established a systematic approach to the selection of appropriate IL and ABS [8]. The ILs' properties that influence the enzyme stabilization and solubilization, namely polarity, anion basicity, IL/water network and viscosity [9] were converted into effects of the medium parameters such as water content (water activity), pH, temperature and type of the system constituents. The strategy followed was first to screen for suitable ILs in highly concentrated aqueous solutions (75% v/v) with respect to the lipolytic activity remained and the enzyme conformation retained, then to select the other phase-forming constituent (type and concentration), and afterwards to construct a successful ABS for efficient and safe enzyme extraction.

It is well known that water plays a crucial role in non-conventional (predominantly non-aqueous) biocatalytic media [7]. Most biocatalysts need a certain amount of water even though the quantity required could seem surprisingly small [10]. This water is directly bound to the enzyme molecule. The water present in a solution could hydrate the other compounds but without contributing to the biocatalyst's activity. This is the reason why all the three main constituents of an ABS should be balanced in order to fit best the enzyme. For example, the ABS composed of $[C_2C_1im][C_2SO_4]/K_2CO_3$ has the best extractability in the IL-rich phase at 13.1/25.0% (w/w) (Table 7.1.1). The deviation toward the higher IL concentration results in decreased efficiency as less free water remains available to hydrate the enzyme. The extractability of the same IL-family based on alkylsulfate anion was also studied with lipase enzyme from a different producer (yeast *Candida antarctica*) was worked out [11]. The ILs of moderate alkyl chain length of the anion inhibit less the lipolytic activity which was explained by the optimal polar network created around the enzyme. *Candida antarctica* lipase A suffers from drastic inactivation by the most employed inorganic salts in ABS, K_3PO_4 and K_2CO_3, due to their strong interactions with the enzyme moiety. For this reason, $(NH_4)_2SO_4$, being a common agent in protein purification methods which exhibits a moderate salting-out ability, was used to design an ABS. The results reported about the lipase extraction into the IL-rich phase show extremely high distribution coefficients which mean that the enzyme could be recovered ultra-concentrated by applying ABS in a wide range of IL-to-salt ratios (see Table 7.1).

Another IL/salt ABS comprising common and cheap constituents, $[C_2C_1im]Br$ and potassium phosphate was engaged in recovery and purification of lipase from fermentation broth of recombinant *E. coli* [12]. Under the optimization of the process parameters including ABS composition, addition of NaCl and crude enzyme loading, lipase was successfully recovered (yield of 94%) and purified (3.4-fold) in a single extraction step (Table 7.1).

Table 7.1: Overview of the ABS with ILs applied for lipase recovery and purification.

Lipase enzyme (characteristics)	ABS The IL	composition The other	ABS pH/other adds	Extraction efficiency	Purification factor	Reference
Lipase from fungi *Thermomyces lanuginosus*, ~33 kDa, pI ~4.4	$[C_2C_1im][C_2SO_4]$, 13.1% w/w	K_2CO_3, 25.0% w/w	Highly alkaline pH	~80% of the initial enzyme with 99% of activity	No data	[8]
Lipase A from yeast *Candida antarctica*, ~45 kDa, pI ~7.5	$[C_2C_1im][C_4SO_4]$, 9.2% w/w	$(NH_4)_2SO_4$, 32.2% w/w	pH ~ 5–6	~99% of the initial enzyme	Partition coefficient of 1639	[11]
Lipase from bacterium *E. coli* fermentation broth	$[C_2C_1im]Br$, 26.5% w/w	K_2HPO_4/KH_2PO_4, 19.0% w/w	pH = 7.6 3.0% NaCl; 7.0% crude load	93.75% of the initial enzyme;	3.4-fold purification in the IL-rich phase	[12]
Lipase B from yeast *Candida antarctica*, 35.3 kDa, pI ~ 6.0	$[C_8C_1im]Cl$, 25.0% w/w	K_2HPO_4/KH_2PO_4, 30.0% w/w	pH = 7.0	~96% of initial lipase recovered in salt-rich phase	2.6-fold purification in salt-rich phase	[13]
Lipase B from yeast *Candida antarctica*, 35.3 kDa, pI ~ 6.0	$[C_6C_1im]Cl$, 5% w/w adjuvant to PEG 1500, 15% w/w	K_2H/KH_2PO_4, 15.0% w/w	pH = 7.0	~ 52% of initial lipase recovered in salt-rich phase	5.22-fold purification in salt-rich phase	[17]
Lipase from culture broth of *Bacillus* sp., 54 kDa, pI = 3.12	$[C_8C_1im]Cl$ 25.0% w/w	K_2HPO_4/KH_2PO_4, 30.0% w/w	pH = 7.0	92.2% of initial lipase recovered in salt-rich phase	~51-fold purification in salt-rich phase	[14]
Lipase from culture broth of *Bacillus* sp., 54 kDa, pI = 3.12	$[C_6C_1im]Cl$, 5% w/w adjuvant to PEG 1500, 15% w/w	K_2H/KH_2PO_4, 15.0% w/w	pH = 7.0	77.6% of initial lipase recovered in salt-rich phase	103.47-fold purification in salt-rich phase	[17]

Lipase from *Burkholderia cepacia* of commercial purity, ~33 kDa, pI ~ 5	$[N_{4,4,4,4}]$[BES] 27.0 w/w	$K_3C_6H_5O_7$, 20.0% w/w	pH = 7.0	99.2% of initial lipase recovered in the IL-rich (top) phase	3.57-fold purification in the IL-rich (top) phase	[22]
Lipase from *Burkholderia cepacia* of commercial purity, ~33 kDa, pI ~ 5	[Chol][BIT] 35.0 w/w	THF, 40.0% w/w	pH = 3.8	97.61% of initial lipase recovered in the IL-rich (bottom) phase	n.d.	[24]
Lipase from culture broth of *Bacillus* sp. ITP-001, 54 kDa, pI = 3.12	[Chol][BIT] 35.0% w/w	THF 40.0% w/w	pH = 3.8	90.0% of initial lipase recovered in the IL-rich (bottom) phase	136.8-fold purification in the IL-rich (bottom) phase	[24]
Lipase from culture broth of *Burkholderia cepacia* ST8, ~33 kDa, pI ~ 6.3	[Chol][BES] 9.0% w/w	PPG 400 50.0% w/w	pH = 7.0 35% crude enzyme feedstock	99.3% of initial lipase recovered in the IL-rich (bottom) phase	17.96-fold purification in the IL-rich (bottom) phase	[25]

7.1.1.1.2 Lipase recovery in the salt-rich phase of the ABS

Another group of studies dealt with lipase recovery and purification through enzyme isolation in the salt-rich phase of the ABS [13, 14]. First multiple ILs were combined with K_2HPO_4/KH_2PO_4 to induce formation of ABS with pH = 7.0 and they were tested on a model enzyme of *Candida antarctica* B lipase type. An additional 2.6-fold purification (over the commercial grade) was achieved by applying $[C_8C_1im]Cl$ (Table 7.1). This ABS offers several conveniences for protein partitioning that include (1) stable phases formed and equal system volume distribution between them; (2) nearly full segregation of contaminant proteins in the IL-rich phase; (3) isolation of the target protein in the salt-rich phase. The lipase partitioning in the salt-rich phase is probably an electrostatic event related to the electrostatic state of the enzyme molecule at pH = 7.0. As this enzyme has pI ~ 6.0, it is slightly electronegatively charged at pH = 7.0, which means that it is likely to be attracted by the more charged phase (the saturated solution of the inorganic salts) than the aqueous solution of the low dielectric IL's ions. This ABS was promoted to purification stage in a real process of enzyme production, submerged fermentation of *Bacillus* sp. ITP-001 to produce extracellular lipolytic enzyme [14]. The whole process chain is schematically represented in Figure 7.1 and a detailed information about the enzyme activity on the process route is given. Some pre-purification steps were involved, including (1) precipitation of contaminant proteins with $(NH_4)_2SO_4$, followed by (2) dialysis aiming at removing the low molecular weight compounds. After that extractions by ABS with Cl^- and $[N(CN)_2]^-$ -based ILs conjugated with potassium phosphate buffer were carried out under the conditions of the previous optimization report. All the systems studied gained significant purification but the best result, purification factor of 51, was obtained with

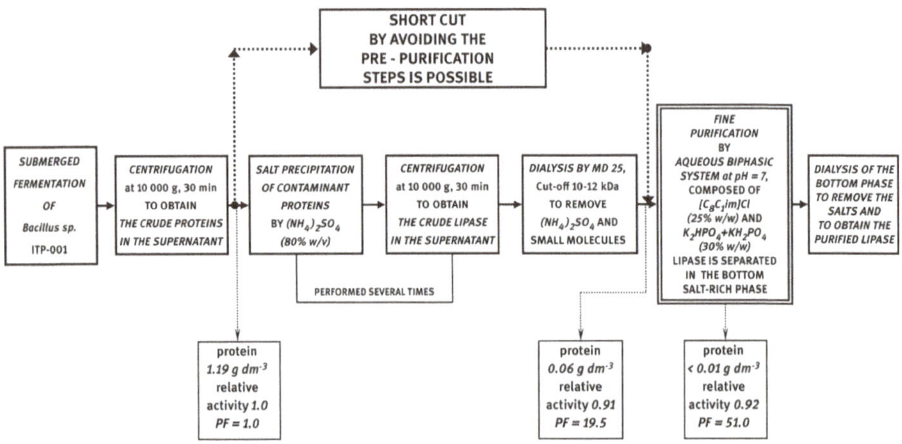

Figure 7.1: Technological scheme for purification of microbial lipase by IL-based ABS (*acc. to* the data reported in [14]).

ABS of $[C_8C_1im]Cl$. Obviously this system proposed the most favorable conditions for accommodation of the contaminant proteins in the IL-rich phase. Although the nature of these impurities was not discussed, the authors suggested that the differences in partitioning behavior between the contaminant proteins and the lipolytic enzyme must result from the differences in their hydrophobic/hydrophilic character. The lipase enzyme, due to its low isoelectric point (pI = 3.12) [15], is negatively charged at pH = 7.0, which leads to the increase in its hydrophilic affinity for the salt-rich phase. On the other hand, the hydrophobic nature of the $[C_8C_1im]^+$-cation contributes to the decrease in Coulombic interactions and increases the dispersive forces between the hydrophobic proteins and the IL in the IL-rich phase. The proteins that were discarded in the IL-rich phase were of great excess compared with the lipase enzyme, but of similar molecular weight, which makes them difficult to be removed by conventional ABS.

Since the IL-based ABS are used for the purpose of replacing the polymeric ones, a study was carried out in order to compare the factors of purification that could be achieved. The crude culture was first pre-purified (dialyzed) and the dyalisate was used in the ABS. The lipase partitioning in the salt-rich phase of the IL-based ABS resulted in additional 2.45-fold purification. The polyethylene glycol-based ABS showed higher values [16], however, it was clearly shown that the purification by this polymeric system underwent a drastic decrease at pH between 6.0 and 7.0 becoming lower than the purification by the IL. Another advantage of the IL-based ABS over the PEG-based ones could be gained from the lower viscosity of the system. The viscosities of the salt-rich phases of both systems were low and similar, but the polymer-rich phases (of PEG 600 and PEG 4000) showed an order of magnitude higher dynamic viscosity while the IL-rich phases kept viscosities close to the salt-rich phases. This could substantially facilitate the solute transport between the two phases, as well as the system handling as a whole.

7.1.1.1.3 Lipase purification by ABS based on PEG and ILs as adjuvants

Despite the encouraging results obtained by replacing PEGs with ILs, an essential drawback emerges from the ILs' prices which still remain too high to compete with conventional solvents. This pushes the researchers to search for novel approaches which require lower amounts of ILs. Such approach is dealing with ILs as adjuvants to PEG-based ABS. The partitioning and purification of a model lipase B from *Candida antarctica* and a lipase from the culture broth of *Bacillus* sp. ITP-001 were investigated [17]. Series of $[C_nC_1im]Cl$ were studied as 5% (w/w) adjuvants. It was found that the effect on the recovery of contaminant proteins in the PEG-rich phase is related to the alkyl side chain length. $[C_6C_1im]Cl$, which possesses a side chain of moderate hydrophobicity, contributed to the purification. IL with butyl side chain had no effect, while ethyl and octyl ones helped dispersing the protein impurities into the salt-rich phase along with the lipase enzyme.

Different effects were manifested due to the different affinities of these ILs for the PEG-rich phase. It was found that [C_6C_1im]Cl has the highest partition coefficient in the PEG-rich phase [18]. The ILs with longer side chain length undergo autoaggregation by forming micelles [19, 20], thus entraining migration of the solutes toward the salt-rich phase as it was clearly shown for two dyes [18].

Besides the effect of the IL's alkyl side chain length, the polymer molecular weight also plays a role in the removal of the contaminant proteins present in the lipase preparation. Their recovery in the top phase was favored by the short chain PEGs which was explained by the decreased intrinsic viscosity of the solvent. An opposite effect was noticed regarding the isolation of the lipase in the bottom salt-rich phase. Although the enzyme preferentially domains in the lower phase, the combination of high molecular PEG with [C_6C_1im]Cl adjuvant achieves somewhat a better enzyme retention in the salty phase. It could be summarized that low molecular PEGs entrain more protein to be transferred to the top phase, including all contaminant proteins, but also a great part of the enzyme. Based on the above findings, the ABS of PEG + IL adjuvant was applied to purify the lipase produced in a real fermentation broth of *Bacillus* sp. ITP-001. Two process routes were followed. The one included pre-purification treatment by obtaining a dialysate, and the other used clarified culture broth to feed the PEG+ IL-based ABS [17, 21]. In both cases, the [C_6C_1im]Cl adjuvant combined with different PEGs provided the highest purification degrees and a constant yield of ~77–81% lipase in the salt-rich phase (see Table 7.1). In general, the differences in the recovery efficiencies between the PEG-based ABS with and without IL adjuvant were small but they looked remarkable on the scale of purification grade. The approach without pre-purification represents the alternative downstream processing for industrial sectors which do not require a high purity of the outcome enzymes. Isolation of the enzyme and recycling of the ABS components were suggested by applying dialysis of both phases, thus accomplishing the industrial relevance of the process.

7.1.1.1.4 Lipase purification by ABS based on self-buffering ILs

More recently a new type of self-buffering ILs based on Good's buffer anions and [$N_{4,4,4,4}$], [$P_{4,4,4,4}$], or [Chol] cations were proposed as extractants for bacterial lipase from *Burkholderia cepacia* [22]. This kind of ILs comes to replace the phosphate buffer which is the most used salting-out agent in ABS dealing with enzymes that require a controlled pH. It was shown that the incorporation of a bulky and hydrophobic cation into the Good's buffer structure has a pronounced effect on the enzymatic activity by enhancing it up to three times. As already mentioned the activity of the lipase, being an enzyme that naturally acts at the water/oil interface, is strongly dependent on the hydrophilic/lipophilic balance of the solvent. The IL could mediate between the enzyme and the substrate by improving their relationship and intensifying the catalytic action [23]. The potential of Good's buffer ILs to create efficient and sustainable extracting ABS with a biodegradable salt, potassium citrate, was studied [22]. The model enzyme used in partitioning (lipase from *B. cepacia*, ~33 kDa) was found to migrate to the IL-rich (top) phase (see Table 7.1). The authors explained this behavior by surface-dependent phenomenon which is mainly due to the hydrophobic interactions of the enzyme with the hydrophobic groups of the IL's moieties. The electrostatic interactions between the negatively charged amino acids on the surface of the enzyme molecule (at pH = 7.0) and the IL's cation also contribute to the transfer. The highest distribution was achieved by ILs based on quaternary ammonium or phosphonium cations and hydrophilic anions, $[BES]^-$ and $[TAPSO]^-$. These ILs are known to have the capability of self-aggregation in micelles. The micelle formation could be considered the vector that cumulates the electrostatic and hydrophobic interactions between the biomolecule and the IL. Taking advantage of the enhanced lipase activity in the Good's buffer ILs, the ABS can be used as a tool for separating the biocatalyst (lipase) from the reaction products in a hypothetic enzymatic conversion [22]. Coupling reaction and extraction in one ABS is a sustainable trend in downstream processing.

Besides the hydrophobic quaternary ammonium and phosphonium Good's buffer ILs, some organic compounds in high concentrations also positively affect the long-term stability of *Burkholderia cepacia* lipase. This was shown for tetrahydrofuran (THF), an organic solvent often used in extractions from vegetables. THF was combined with [Chol]-based ILs to create ABS for lipase purification [24]. The recovery conditions were optimized in terms of the type of IL's anion and the concentrations of IL and THF. The increase in the IL concentration had effect on the lipase segregation in the bottom (IL-rich) phase of the ABS (Table 7.1). Based on the study of the temperature dependence of the process, it was concluded that the transfer is endothermic and governed mainly by entropic forces. The optimized [Chol][BIT]/THF ABS was applied for purification of the lipolytic enzyme from the dialyzed fermentation broth of *Bacillus* sp. ITP-001 [24]. Recovery of 90% of the active enzyme and 137-fold purification represent the best results ever achieved for IL-assisted enzyme extraction with ABS (see Table 7.1.).

Recently, a biocompatible IL, [Chol][BES], and a thermo-sensitive polymer, PPG 400, were combined to form an ABS that was applied for lipase recovery from *Burkholderia cepacia* fermentation broth [25]. Recovery yield of 99.3% and high purification factor of 18 were achieved after optimizing the system composition and the feedstock loading (Table 7.1). Extractions with recycled phase-formers resulted in the same purification values as those obtained by fresh chemicals. This ABS possesses all the advantageous features of a viable and sustainable platform for lipase purification, being a high biocompatible, marginally toxic, non-volatile, self-buffering and recyclable medium where fast and simple downstream processing can be performed [25].

To summarize, the IL-based ABS proved to be effective means to achieve an additional purification of commercial grade enzymes and can be used as a post-purification technique. On the other hand, the ABS can operate directly with the clarified culture broths thus avoiding the need of pre-purification step applied in most downstream processes. The IL-based ABS developed for separation of lipolytic enzymes and tested on the fermentation broth of *Bacillus* sp. ITP-001 achieved 90% recovery of the active enzyme and purification factors above 50. The most effective ABS are [Chol][BIT]/THF and the PEG/salts ones upgraded by the IL adjuvant. The latter systems are recommended from a practical viewpoint taking into account the price of the ILs which still remains too high. The adjuvants used are selected among the most common, and thus cheap, imidazolium halides, and are applied in small quantities of 5% of the system weight. These additives provide an improvement in the purification potential of the ABS, which can also be applied directly to the clarified broth without pre-purification treatment. In this case, the lipase enzyme was obtained purified in the salt-rich phase and afterwards was isolated by a common dialysis technique. When the lipase was isolated in the IL-rich phase, it was subjected to the same treatment (dialysis) after which the phases' constituents can be re-used.

7.1.1.2 Carbohydrate-converting enzymes

First study that dealt with IL-based ABS and carbohydrases described an IL/salt system for recovery of cyclodextrin glycosyltransferase (CGTase) from fermentation broth [26]. CGTase is an enzyme that catalyzes intramolecular (cyclizing) and intermolecular (coupling, disproportionation) transglycosylation while having a hydrolytic action on starch and cyclodextrins [27]. Previously polymer/salt ABS were applied for CGTase recovery and successful (13–16-fold) purification was achieved [28, 29]. Considering the long settling time and the centrifugation required to attain the phase separation of the polymeric systems which are disadvantageous for large-scale production, as well as the polymer recycling which is often complicated, the polymer constituent was replaced by an IL in order to improve the cost-efficiency. The ABS was composed of [C_2C_1im][BF_4] and sodium carbonate or citrate. First the preservation of the CGTase activity was studied in terms of 1-hour incubation in aqueous solutions of the phase-forming components, the IL and the salt, at varying concentrations. It was shown that the activity was retained by using less than 40% (w/w) of both the IL and the salts. Deterioration of the activity at the higher IL and salt concentrations was attributed to the decrease in the number of free water molecules available to hydrate the enzyme. The IL strips the essential water from the biomolecule thus causing structural (conformational) alterations, whereas in the salty solutions the CGTase is salted-out and its precipitation was observed. The IL/salt ABS suitable for CGTase recovery was created with low concentrations of [C_2C_1im][BF_4] and sodium salt. Na_2CO_3 provided higher distribution of the enzyme than $Na_3C_6H_5O_7$. Migration of the CGTase toward the IL-rich phase could be explained by electrostatic interactions between the enzyme and the system constituents. This ABS possesses alkaline pH (= 9.38) which is above the CGTase isoelectric point (pI = 6.9) so the negative charges predominate on the enzyme molecule surface. The strong repulsion exerted by the salt-rich phase (divalent anions) on the electronegatively charged enzyme causes its migration toward the IL-rich phase. The transfer is also facilitated by the electrostatic attraction of the IL's cation, $[C_2C_1im]^+$, and the inert IL's anion, $[BF_4]^-$. The purity of the CGTase extracted into the IL-rich phase was strongly dependent on the small variations, within 3–5%, in the concentrations of both phase-forming compounds. The best results were obtained by using composition of 35% $[C_2C_1im][BF_4]$ and 18% Na_2CO_3 (Table 7.2). Both the IL and the salt finely tune the hydrophilic–hydrophobic balance of the system. By increasing their concentrations the hydrophobicity is increased. The CGTase is repelled from the salt-rich phase through salting-out effect and transferred into the IL-rich phase with the aid of hydrophobic interactions with the IL. At higher concentrations of both the IL and the salt protein precipitation at the interface was observed due to enzyme dehydration. The initial protein loading also influenced the purification efficiency and the extraction yield. The lower loadings of the crude enzyme feedstock were found more favorable, 15% w/w were optimal for the selected ABS composition. The higher loadings were associated with high percentages of impurities. Impurities might alter the electrostatic potential of the ABS as they react with the phase components through electrostatic interactions. The excluded-volume effect may also interfere and cause reduction in the CGTase recovery. The purification extent and the yield could additionally be enhanced by introducing an electrolyte. The addition of 3% w/w NaCl increased the yield by more 6%, up to 96.23%, and the purification factor increased respectively (Table 7.2). The influence of the NaCl addition on the phases' composition was inferred from the phases' inversion observed due to the change in the phases' densities.

The $[BF_4]$-based IL was used as adjuvant to boost the yield of β-mannanase extracted from *Bacillus subtilis* ATCC 11,774 fermentation broth by $PEG/K_3C_6H_5O_7$ ABS [30]. Although a limited range of PEGs was studied it was found that those with higher weights (PEG 6000) better fit the mannanase activity. The pH = 8.0 was the most appropriate for the enzyme partitioning and purification in the PEG-rich (top) phase. Distribution was also favored by low loading of crude enzyme. The addition of hydrophilic ILs ($[C_2C_1im]Br/[BF_4]$) as adjuvants was found to negatively affect the partitioning and purification as these ILs probably partition into the salt-rich (bottom) phase and entrain the enzyme toward it. The addition of $[C_4C_1im][BF_4]$ insignificantly changed the distribution and purification parameters, however, it caused a notable improvement in the recovery yield of the target enzyme from 62.25 to 89.65%, an increase of 27% (Table 7.2).

On the other hand, PEGs could be applied as adjuvants to the IL/salt ABS. This possibility has been recently addressed to the partitioning of the α-amylase enzyme in $[C_4C_1im]Ac/K_2HPO_4$ ABS [31]. Based on calculations by response surface methodology, the effects of the phase-forming components on the partition

Table 7.2: Overview of the ABS with ILs applied for carbohydrase recovery and purification.

Carbohydrate-converting enzyme	ABS The IL	Composition The other	ABS pH/other adds	Extraction efficiency	Purification factor	Reference
CGTase from bacteria *Bacillus cereus*; 55 kDa, pI = 6.9	$[C_2C_1im][BF_4]$, 35.0% w/w	Na_2CO_3, 18.0% w/w	pH = 9.38 3% w/w NaCl 15% crude load	96.23% of the initial enzyme	13.86-fold in the IL-rich phase	[26]
β-Mannanase from bacteria *Bacillus subtilis* ATCC 11,774 36.7 kDa, pI = 4.5	$[C_4C_1im][BF_4]$, 3% w/w adjuvant to PEG 6000, 25% w/w	$K_3C_6H_5O_7$, 12.52% w/w	pH = 8.0 1.5% crude enzyme feedstock	89.65% of initial mannanase recovered in PEG-rich (top) phase	~3-fold purification in the top phase; $K_p \sim 1$	[30]
Commercial product α-amylase from *B. licheniformis* 62.0 kDa, pI = 5.25	$[C_8C_1im][Sac]$, 37.7% w/w (33.3% v/v)	Aqueous buffer, 0.150 mol dm^{-3} (66.7% v/v)	pH-range of extraction ~4.5–10.5	At pH ≤ pI, 54% in the re-extract; At pH > pI, 99% in the raffinate	1.2–1.3	[33]

K_p: partition coefficient.

coefficient of α-amylase were evaluated. Taking into account the effective excluded-volume theory that correlates the salting-out ability, it was stated that an increase in the initial concentration of the inorganic salt makes the space of the bottom phase more packed and thus the enzyme is pushed to the top phase. The nature of [C_4C_1im]Ac was considered mainly hydrophobic based on the data reported about the hydrogen bond basicity of 1.2 [32]. The addition of PEG reduces the phases' miscibility as the mixtures of IL and PEG are more hydrophobic than the pure counterparts and are more easily salted-out by K_2HPO_4. It was concluded that the extraction process is mainly governed by the hydrophobic interactions and the size-exclusion effect. At higher concentrations or at high molecular weights, PEG molecules may exclude the biomolecules to the lower salt-rich phase. Under the proper selection of the process variables (PEG weight and concentration) the solute partition coefficient could be optimized. Since these conclusions were based on a mathematical study by statistical experimental design, they need some validations by material experiments.

Recently an experimental study revealed the opportunity to control the partitioning of α-amylase enzyme in ABS of a hydrophobic and polar IL, [C_8C_1im][Sac] [33]. The imidazolium saccharinates with long side chain, [$C_{8/10}C_1$im][Sac], are able to form ABS alone without the aid of another phase-separating compound. This occurs just after the ILs reach saturation of about 55% w/w of water. Stable biphasic region on the phases' diagram exists between 45 and 5% w/w of the ILs [34]. Stability and activity of α-amylase enzyme (from *Bacillus licheniformis*) in the presence of different concentrations of the IL were studied. It was found that the activity deficiencies occur at the high IL concentrations. The loss of activity was more pronounced when α-amylase was set at pH below pI. The protonated carboxyl groups and positively charged amine groups of the amino acids dominate on the biomolecule surface at pH < pI which suppose that the IL interacts with this kind of enzyme form and these interactions are responsible for the loss of activity. It was found, however, that the deactivation was reversible and the activity could be restored in the presence of acetate anions. Based on this an ABS of [C_8C_1im][Sac]/acetate buffer was developed which enabled controlled partitioning of the α-amylase to be achieved by pH variation below and above the enzyme pI (Table 7.2).

At pH ≤ pI 54% of the initial α-amylase enzyme was transferred through the IL-rich phase into another aqueous solution. This re-extract was purified from the inactive (as α-amylase) protein and was substantially discolored from the brown pigment produced in *B. licheniformis* fermentation during the long culture period [35]. On the other hand, under the conditions of pH > pI the enzyme transfer into the IL-rich phase was entirely restrained and the α-amylase remained in the raffinate. Some contaminant proteins, however, were transferred. The selectivity achieved in the extraction–re-extraction cycle (re-extracts and raffinates) by changing the source pH below and above the pI of the α-amylase is illustrated in

Figure 7.2. Data about selectivity of the reversed micelle extraction method applied for purification of the same enzyme source are also presented [36]. Both methods offer the opportunity to achieve selectivity in either phase depending on pH. Similar trend can be observed in α-amylase partitioning in PEG/potassium phosphate ABS by varying the PEG's weight [37].

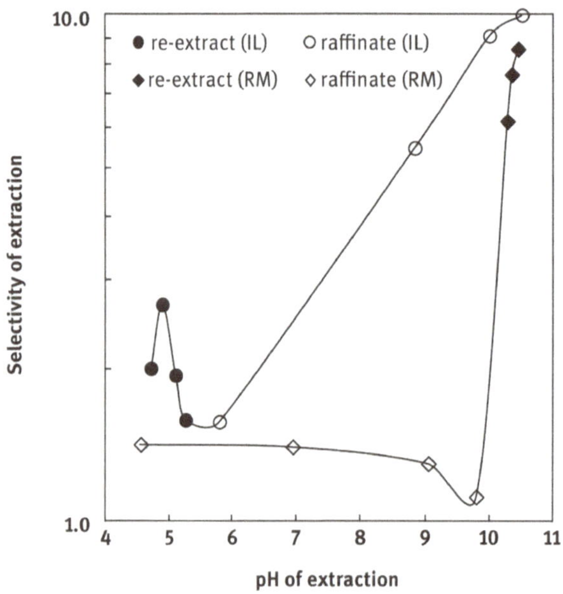

Figure 7.2: Selectivity of extraction of α-amylase enzyme (from *B. licheniformis*) using IL [33] or reversed micelles, RM [36].Selectivity is determined as the ratio between the specific α-amylase activities in the aqueous phases after forward and backward extractions.

The purified α-amylase in the raffinates corresponded to *ca*. 75% of the whole protein. The share of the inactive (as α-amylase) protein, 25%, resembled the findings of the reversed micelle extraction of the same α-amylase product [38]. Similarly the α-amylase extraction with IL was favored by the low loadings of the enzyme preparation [39]. The reversed micelle extraction, however, is strongly dependent on the addition of electrolytes (inorganic salts like Na/K/NH_4–Cl/Br) in both aqueous phases, the source and the stripping solution [7, 36, 38]. In small concentrations they can significantly boost the degree of purification. Such effects have not been studied yet regarding the IL-based systems, except for the small increase in the yield of CGTase extracted in the IL-rich phase by the addition of NaCl [26].

The factors and parameters that affect the enzyme partitioning in the IL-based ABS are discussed next.

7.1.2 Factors and parameters affecting the partitioning of enzymes in ABS with ILs

Besides the electrostatic interactions between the enzyme and the IL and the salting-out effect of the salt, which both are considered driving forces of the extraction, the secondary macrostructure of the IL-rich phase and the water content are also responsible for the preservation of the enzyme structure and activity during the transfer process.

7.1.2.1 Choice of IL and pH. Interactions between IL and enzyme

The choice of IL and salt components, the ions' size and structure might be decisive factors that affect the protein conformation in the ABS. Sterically demanding ions would require many hydrogen bonds to be broken in order to create fewer ones, which could disrupt the protein hydration [11]. IL's and salt's ions are able to alter the bulk water structure and thus to affect the protein–water interactions or directly interplay with the enzyme molecule modifying its affinity for the substrate. As a rule, the choice of appropriate IL and salt constituents of an ABS must be founded on investigation of enzyme deactivation profiles in the presence of different IL and salt concentrations. Many of the published results revealed the relationship between the enzyme activity and the hydrophilic–lipophilic balance of the enzyme-solvent system. The hydration level at the enzyme surface could be regulated through the nature of the IL's ions. It was shown that unlike the cholinium cation, the ammonium and phosphonium cations comprising hydrophobic alkyl chain do not strip the hydration shells off the enzyme surface [22]. In the IL/salt, ABS the protein/enzyme migration to the IL-rich phase is enforced in two manners: electrostatically, by a strong repulsion exerted by the salt-rich phase, which usually has pH above the protein pI [26, 40] or as a result of a surface phenomenon, the formation of IL aggregates-protein complexes [41, 42]. Regarding the ILs that exhibit surfactant properties, the interfacial transfer, mainly driven by hydrophobic interactions, can be assisted by electrostatic attraction between the IL's cation and the negatively charged aminoacid residues. These systems work as IL-supported aqueous micellar two-phase systems [43]. The IL $[C_{10}C_1im]Cl$ has shown an excellent cell lysis ability which enables the release of the intracellular and membrane-bound enzymes that are afterwards separated in the IL-rich phase of the ABS $[C_{10}C_1im]Cl/(NH_4)_2SO_4$ [44].

The ABS based on the most common IL with surfactant properties, $[C_8C_1im]Cl$, were found unable to extract the enzyme (lipase) into the IL-rich phase but able to host the contaminant proteins thus leaving the purified enzyme in the salt-rich phase [13, 14]. Another promising technique is based on the use of ILs as adjuvants to polymeric ABS. The enhanced extraction of impurity proteins into the (PEG+ IL)-rich phase has led to utmost factors of lipase purification in the salt-rich phase. This was attributed to the synergetic effect between the IL and the PEG [17]. Elongation of the alkyl side chain length on the imidazolium ring has a significant effect on the partitioning of the target and contaminant proteins. Distribution coefficients of the ILs increase from C_2 to C_6 [18] enhancing the positive charge of the PEG-rich phase. This results in mutual transfer of enzyme and contaminant proteins to the (PEG + IL)-rich phase. It should be noted that the enzymes studied (lipases CaLB and *Bacillus* sp. ITP-001) are negatively charged at the given pH (see Table 7.1). Involving IL of C_8-chain pushes back the transfer of both the enzyme and the impurities toward the salt-rich phase. This is explained by the micelle formation in the bottom phase (lower partition coefficient of $[C_8C_1im]Cl$ compared to $[C_6C_1im]Cl$). The negative effect of the presence of micelles on the transfer toward the IL-rich phase was also described for $[C_nC_1im]Cl$/salt ABS [45]. The addition of $[C_8C_1im]Cl$ to the PEGs causes the lowest distribution of proteins in the top phase, especially with high weight PEGs when the intrinsic viscosity of the phase increases. In general, the choice of an IL as adjuvant to polymeric ABS should be made regarding the IL's chemical structure as it alters the hydrophilic/hydrophobic nature of the polymer.

Electrostatic interactions between the IL's cation and the negatively charged enzymes are responsible for the recovery in the top (IL or PEG-rich) phase. This is described for ABS with ILs composed of the $[C_{2/4}C_1im]^+$ cation and the inert $[BF_4]^-$ anion (see Table 7.2). Regarding the selection of IL and IL's structure it can be summarized that the short side chain imidazolium ILs mainly act by exerting positive electrostatic potential toward the negatively charged enzymes. Long side chain imidazolium ILs with tensioactive nature, which are capable of auto-aggregating in micelles, do not support the transfer of the enzyme into the IL or PEG-rich phase thus promoting its separation into the salt-rich phase.

Some studies showed that the ILs can extract the enzymes at pH below the pI-values probably by establishing H-bonds between the polarized IL–H_2O complexes and the protonated carboxylic groups on the amino acids' chain [11, 33]. Though this results in enzyme deactivation [8, 33], the process is often reversible [33]. This kind of interactions can be used to drive the extraction with caution against the unfolding and deactivation.

It is worthwhile to carry out more studies with hydrophobic (quaternary ammonium, phosphonium) ILs as phase-forming components or adjuvants. It has been already shown that the quaternary ammonium ILs extract the lipase enzyme mostly by hydrophobic interactions through surface-dependent mechanism [22]. The application of ILs as adjuvants in small amounts is currently more profitable regarding the high price of most ILs.

7.1.2.2 Water content and structure of the water saturated IL-rich phase. Effects on the enzyme activity

As already mentioned, water is the key component of all kinds of biocatalytic media, especially the non-conventional (predominantly non-aqueous) ones [7, 10].

Most biocatalysts require plenty of water though for some only small amount is enough. The water in an ABS attends hydration of the phase compounds and the free water that remains may be insufficient to support the proper enzyme conformation for the biocatalytic action. For this reason, the concentrations of the IL and the salt have to be experimentally determined in terms of defined enzyme/total protein loadings. Generally, the ABS do not operate well at higher protein concentrations due to the risks of biomolecules aggregation under the conditions of restricted water content. Related to the water network, the ion size of the IL can influence the enzyme activity, since the sterically demanding ions could disrupt the protein hydration [46]. IL's and salt's ions are able to alter the bulk water structure and thus to affect the protein–water interactions or directly interplay with the enzyme molecule modifying its affinity for the substrate.

The specific ion effects associated with the water content can favor the enzymatic action or can disturb it. It has been recently shown that the α-amylase enzyme could be preserved in full activity for a long grace period (up to two weeks) in highly concentrated aqueous solutions of $[C_4C_1im]Br$, 3.66 mol dm^{-3} (80.2% w/v) [47]. The incubation of the enzyme in this IL's solution brought about a reduction of 50% in activity for five hours. It was found, however, that a trace amount of an acetate salt in this $[C_4C_1im]Br$ concentrated aqueous solution had a huge effect on the long-term activity preservation. The addition of the acetate salt, even in minor concentration (0.05 mol dm^{-3}) compared to the IL, substantially altered the polar network surrounding the enzyme and changed the pH value of the solution. These two effects were beneficial to the activity retention and it was optimized though the salt type and concentration. The best results were obtained by adding potassium acetate, 0.20 mol dm^{-3}. Some 90% of the activity was saved in the two-week incubation, and a half activity remained in a month. The grace period of the preservation of full activity, however, was drastically shortened when the salts were added in high concentrations. Increasing the salt amount more water molecules became involved in the solubilization of the salt and less free water remained to hydrate the enzyme. Salts with high charge density, for example K_2HPO_4 which has strong salting-out ability, severely deactivated the enzyme.

The water activity emerges as a key factor of the utmost importance in controlling the enzyme activity in the ABS. Taking into account the linear relationship between the water activity and the water content of the solvent [10], the latter can be considered the key parameter in maintaining the enzyme conformation intact and keeping the catalytic activity [39].

7.1.2.3 Choice of in/organic salt. Salting-out effect

Most of the studied IL-based ABS are formed with kosmotropic salts comprising inorganic or organic anions. The salting-out exerted over both the IL and the enzyme is considered an important factor in enzyme partitioning according to the published literature. As already mentioned, the choice of phase-segregating promoter is principally based on the investigation of the enzyme deactivation. This should be carried out case-by-case for each individual enzyme. As an example, the lipase enzyme from fungi is tolerant to the high concentrations of K_2CO_3 [8], but *Candida antarctica* lipase A is unfolded and deactivated even after a short contact time [11]. In many cases phosphate salts, K_2HPO_4/KH_2PO_4, are preferred as they allow pH adjustment or $(NH_4)_2SO_4$ is used, which is a salt widely applied for protein precipitation, one of the common protein purification methods.

A few studies dealing with the ABS of IL/organic solvent demonstrate that the salting-in ability of the IL is the driving force for the enzyme migration toward the IL-rich phase [24]. The salting-in effect of the IL and the salting-out effect of the salt both support the transfer toward the IL-rich phase as this was shown in CGTase extraction [26]. Salting effects can be strengthened by increasing the concentrations of the phase-segregating components, but this is limited to the border of protein precipitation.

7.2 IL-assisted recovery of fermentatively derived organic acids

7.2.1 Overview

Considering the benefits that arise from the properties of ILs, Matsumoto and coworkers [48] first proposed an environmentally friendly system for the extraction of fermentative L-lactic acid. They used hydrophobic $[C_nC_1im][PF_6]$ instead of volatile organic solvents as diluents of reactive organic bases. These ILs proved to be nontoxic toward the lactic acid producing bacterium *Lactobacillus rhamnosus*, but provided low degrees of solubility of the reactive amines which resulted in insufficient levels of extraction efficiency. Nevertheless, these results suggest possible applications of ILs in extractive fermentations.

The next attempts to recover organic acids from their aqueous sources by diverse ILs have been recently discussed in detail [49]. It should be noticed that with a few exceptions, hydrophobic ILs are used that form classical biphasic systems of IL/water. Hydrophobic ILs are hygroscopic and absorb water from air or dissolve water when placed in contact with aqueous solutions. In many cases, due to the great water uptake in the IL-rich phase [34, 50–52] these biphasic systems resemble ABS. Here the information about the organic acids extracted and the ILs applied is summarized in Table 7.3. In addition to the quantitative results about the extraction efficiencies and partition coefficients, the information given is extended over the organic acid characteristics, namely dissociation constant, pK_a, and pH value of the acid aqueous source (used with or without additives), which both affect the acid–IL interactions, as it is discussed next.

Table 7.3: Overview of the ILs applied and the organic acids recovered.

IL	Organic acid pK_a	Organic acid Concentration (M)	Aqueous solution pH	Extraction efficiency	Partition coefficient	Reference
[P$_{6,6,6,14}$][Phos]	Butyric acid	~ 0.27	~2.92	n.d.	~80.00	[52, 53]
[N$_{1,n,n,n}$][Phos]	pK_a = 4.82					
[P$_{6,6,6,14}$]Cl	L-malic acid	0.20	~2.00	75.3%	3.00	[54]
[P$_{6,6,6,14}$][Phos]	pK_{a1} = 3.46	0.20	~2.00	61.1%	3.90	
[P$_{6,6,6,14}$][Dec]		0.05	~2.50	80.0%	4.00	
[P$_{6,6,6,14}$]Cl	Succinic acid	0.20	~2.70	85.9%	6.10	[54]
[P$_{6,6,6,14}$][Phos]	pK_{a1} = 4.21	0.20	~2.70	78.0%	3.50	
[P$_{6,6,6,14}$][Dec]		0.05	~3.00	80.8%	4.20	
[C$_6$C$_1$im]Br/	Succinic acid	~0.17	3.43*	85.5%	1.06	[55]
$(NH_4)_2SO_4$ or	pK_{a1} = 4.21		10.50*	82.8%	2.07	
K_2CO_3	pK_{a2} = 5.72					
[P$_{6,6,6,14}$]Cl	L-lactic acid	0.50	~2.05	61.8%	1.60	[54]
[P$_{6,6,6,14}$][Dec]	pK_a = 3.86	0.50	~2.05	65.7%	1.90	
[P$_{6,6,6,14}$][Phos]		0.50	~2.05	83.3%	5.00	
[P$_{6,6,6,14}$]Cl +	L-lactic acid	0.50	1.88	82.9%	3.38	[56]
+ $MgSO_4$	pK_a = 3.86					
[C$_8$C$_1$im][PF$_6$]	L-lactic acid	0.20	0.90	42.0%	0.72	[57]
[C$_8$C$_1$im][BF$_4$]	pK_a = 3.86	0.20	0.90	69.8%	2.31	
[C$_6$C$_1$im][HClO$_4$]		0.20	0.90	45.0%	0.82	
[C$_8$C$_1$im][PF$_6$]	Citric acid	0.20	0.90	44.3%	0.80	[57]
[C$_8$C$_1$im][BF$_4$]	pK_a = 2.79	0.20	0.90	43.9%	0.78	
[C$_6$C$_1$im][HClO$_4$]		0.20	0.90	43.0%	0.75	
[C$_8$C$_1$im][PF$_6$]	Mevalonic	1.20	0.90	49.1%	0.96	[57]
[C$_8$C$_1$im][BF$_4$]	acid	1.20	0.90	65.2%	1.87	
[C$_6$C$_1$im][HClO$_4$]	pK_a = 4.24	1.20	0.90	68.7%	2.21	
[C$_8$C$_1$im][Sac]	L-lactic acid	0.50	2.04	65.4%	1.06	[34]
[C$_8$C$_1$im][Sac] +	pK_a = 3.86	0.50	1.51	80.5%	5.50	
+ $MgSO_4$						
[P$_{6,6,6,14}$]Cl	Acetic acid	0.17	2.80	n.d.	3.70 1	[58]
[P$_{6,6,6,14}$][N(CN)$_2$]	pK_a = 4.76	0.17	2.80	n.d.	1.80 1	
[P$_{6,6,6,14}$][Phos]		0.17	2.80	n.d.	17.00 2	
[P$_{6,6,6,14}$]Cl	Acetic acid	0.33	~2.00	n.d.	1.71 3	[59]
[P$_{6,6,6,14}$][N(CN)$_2$]	pK_a = 4.76	0.33	~2.00	n.d.	3.05 3	
[P$_{6,6,6,14}$][Tf$_2$N]		0.33	~2.00	n.d.	0.23 3	

n.d.: no data.

*pH of extraction.

[1] The distribution coefficient is not affected by the presence of inorganic salts (KCl, Na_2SO_4 or Na_2HPO_4) in the feed.

[2] The distribution coefficient decreases from 17.0 to 3.3–1.2 when inorganic salts (KCl, Na_2SO_4 or Na_2HPO_4) are added to the feed.

[3] The extraction was carried out at 75°C and the ratio of aqueous phase to IL was 10.

7.2.2 Factors and parameters affecting the extraction of organic acids by ILs and unraveling the extraction mechanism

7.2.2.1 Choice of IL and pH. H-bonding. Reactive mechanism

Although the nature of the IL-driven processes is still unclear and problematic, some general trends could be retrieved from the experimental data hitherto presented in the literature. First of all it is observed that the pH values lower than the acid pK_a mostly favor the extraction. This means that the protonated acid preferably interacts with the IL. Reactive mechanism *via* formation of acid–IL H-bonded complexes has been proved for the extractions of butyric and lactic acids by [Phos]-based ILs [51–53, 60]. Reversed micelles, that exist in the water saturated [$P_{6,6,6,14}$][Phos] and [$N_{1,n,n,n}$][Phos], do not serve the acid transfer. By increasing the acid concentration, the IL's molecules become involved in complexation with acid molecules, which splits the reversed micelles and the water from their cores is liberated. Stoichiometric complexes of multiple butyric acid molecules per one IL molecule prevailed in a broad range of acid concentrations. This suggests that the extracted acid molecules are possibly organized in chains with H-bonds on carboxylic groups. Stability of the H-bond between acid molecules in the chain is relatively low [51, 61]. Therefore, mobility of individual acid molecules, hopping acid molecules, between these chains could be suggested. Maximal loading of IL with some seven acid molecules was achieved in the experiments [52]. Few water molecules were found to constantly hydrate the IL–acid complexes carrying a dimer or multiple acid molecules. This indicates that reversed micelles no longer exist in the IL phase of high acid loading and the water is entirely involved in the complexes' hydration. Equimolar acid–IL complexes exist only in a narrow range of acid concentrations between 0.006 and 0.010 mol dm^{-3}.

Little particularities related to acid's nature can be differentiated in extractions by [$P_{6,6,6,14}$][Phos]. Regarding the lactic acid recovery, the contribution of the physical extraction (multiple molecules extracted) is less pronounced. Equimolar complexes of acid with IL exist in a wider range of acid concentrations, the maximal loading is 3, and only two water molecules hydrate one complex [60].

Another IL that was found to extract the organic acid though H-bonding (reactive) mechanism is [$P_{6,6,6,14}$][$N(CN)_2$] [58]. The extraction efficiency is high at low pH and drops with increasing the pH along with the shift in the acid equilibrium to dissociation. It is noteworthy that an addition of a strong mineral acid in order to lower the pH is not generally recommended. Similar to TOA/n-octanol extractant, [$P_{6,6,6,14}$][Phos] undergoes interfacial protonation, induced by the dissociated protons of the mineral acid, which entrains extraction of counter-anions to maintain the neutrality. This significantly reduces the IL's extractant capacity for the organic acid. [$P_{6,6,6,14}$][$N(CN)_2$], however, makes an exception. The distribution coefficient of the organic acid remains the same in the presence of 1 mol dm^{-3} of HCl or HNO_3. Most likely, protonation of [$P_{6,6,6,14}$][$N(CN)_2$] does not occur, which may also explain the low co-transfer of other anions originating from the salts present in the acid source [58].

All said above reveals that the basicity of the IL's anion plays a significant role in the extraction performance. A guideline for choice of IL extractant is the pK_a value of the conjugated acid of the IL's anion. It is shown that the ILs containing anions of weak acids exhibit good extraction capacity while those of conjugated stranger acids exhibit poor extracting ability [51]. As a rule, the ILs with hydrophobic anion extract only organic acids in undissociated form by H-bonds while the pH of the source should be well below pK_a of the acid to achieve high extraction efficiency. Attachment of the protonated acid by H-bond to the binding sites in the IL's anion (ex. oxygen of [Phos] or carboxylate anion) is supposed to be a preferred mechanism of extraction.

7.2.2.2 Water content and structure of the water saturated IL. Physical extraction

Water uptake of the IL is usually regarded as a feature of the IL's anion H-bonding ability. Water accumulates close to the anion and can form clusters and water network even appears to be formed at higher water concentrations [62, 63]. Aggregates of water formed in [$P_{6,6,6,14}$][Phos] were indirectly detected by the significant change in the chemical shift of the anion, due to the reorganization of its surrounding, that was recorded on the ^{31}P-NMR spectrum [64]. [$N_{1,n,n,n}$][Phos], which absorbs high amount of water (21% w/w), tends to form a phase of supramolecular structure with nonpolar domains and polar channels containing water [52]. The higher-order mesophase contributes to the co-extraction of multiple acid molecules [61].

Long side chain imidazolium saccharinates, [$C_{8/10}C_1$im][Sac], were found able to swell around 55% w/w of water at saturation [34]. It was assumed that nonpolar domains, separated by polar channels capable of swelling and liberating water and other solutes are formed. The order in these channels was deduced on the basis of ^{1}H-NMR measurements. In the proton spectra of [C_8C_1im][Sac] taken in water (D_2O), the chemical shift for the imidazolium proton H-C(2) was found 1.2 ppm downfielded relative to its position in chloroform ($CDCl_3$). This indicates water induced disruption of the H-bond between the IL ions (pronounced in the nonpolar aprotic $CDCl_3$) and suggests that the polar channels are composed of polarized water molecules which interact independently with the polar head of the cation and the saccharinate anion. Thus, the molecule of the protonated organic acid can be considered as a guest molecule capable of inserting into the H-bond embedded polar domains. In contrast to the reactive mechanism of direct acid–IL complexation, discussed above, the case of recovery by imidazolium saccharinates could be described as physical extraction mainly driven by the concentration gradient. It was assumed that the acid molecules are solvated by the water present in the polar channels of the IL-rich phase, rather than directly bound to the IL's ions. Formation of complexes "acid–H_2O–IL" instead of "H_2O–acid–IL" as in the case of [Phos]-based ILs was proposed.

7.2.2.3 Ion exchange between IL and organic acid

Hydrophobic ILs that contain water soluble anions can extract organic acids through ion exchange mechanism. Most studies dealing with Cl-based ILs suggest that ion exchange is one of the extraction mechanisms. This statement was supported by the measurable leaching of chloride anions from the IL-rich phase into the aqueous raffinate [52, 58]. The release of Cl^- results in lower pH of the raffinate compared with the aqueous source. Reyhanitash et al. [58] calculated that the half of the acetic acid extracted in $[P_{6,6,6,14}]Cl$ was due to ion exchange though which Cl^- was replaced by acetate anion. Extractions of lactic or butyric acids with $[P_{6,6,6,14}]Cl$ or $[N_{1,n,n,n}]Cl$ follow the same mechanism [51, 52]. These ILs extract acids from solutions with pH well above the pK_a, but to some extent they can also extract *via* coordination mechanism at pH below the pK_a [65].

Ion exchange is an irreversible interaction which supposes that the subsequent stripping of the acid from the IL would not be possible without replacing the leached Cl^-. This explains the insufficient yields of the re-extracted acids from the Cl-based ILs using only alkali solutions [54].

7.2.3 Procedures to enhance the extraction efficiency and to intensify the extraction process

7.2.3.1 Multi-stage cross-current liquid–liquid extraction

Alike the enzyme extraction with ILs, the acid extraction efficiency decreases with increasing the acid concentration in the feed [51]. In order to enhance the recovery yield multi-stage liquid–liquid extractions have been performed in majority of studies and two cross-current schemes have been applied on purpose. Following the one, the outcoming raffinate is repetitively contacted with portions of fresh IL. The scheme is applied when a high recovery of the solute from the source is intended to be reached. Thus Tonova et al. achieved 90% removal of the lactic acid from the feed after two steps, and additional 5% in the third stage [34].

Series of successive extractions of L-lactic acid, L-malic acid and succinic acid from their individual aqueous solutions were performed by Oliveira et al. [54]. The overall recovery of each acid by $[P_{6,6,6,14}]$[Dec] in two steps exceeded 90%. Great amount of the acid was retained in a third phase formed after the first extraction. It was, however, successfully transferred into the IL-rich phase in the second step without formation of a new middle phase. Re-extractions were carried out to strip the acids from the IL-rich phases by using pH variation (addition of NaOH). This method arrived at poor yields of L-lactic acid (36%) and succinic acid (46%), and L-malic acid only was fairly re-extracted in mass (73%).

Following the other scheme, the outcoming extractant (IL-rich) phase is repeatedly contacted with portions of fresh aqueous source. The scheme is applied for accumulation of the solute in the IL and for the purpose of studying the IL's extracting capacity. It has been recently shown that the volatile fatty acids can be concentrated ~ 65-fold in $[P_{6,6,6,14}]$[Phos]-solvent phase starting from a fermented wastewater of 1% w/w acids' concentration [58]. Moreover, the IL showed a higher capacity compared to the benchmark solvent, 20% w/w TOA in n-octanol. In addition to this, the system can operate at a lower IL-to-feed ratio. The loaded IL was entirely liberated from the acid by washing with KOH (1 mol dm^{-3}). Otherwise the acid release from the solvent phase is possible without consumption of any extra water or chemical by applying vacuum and higher temperatures [66, 67]. This method is advantageous over the classical stripping with an alkali solution as the product is recovered in concentrated acid form, not the salt.

7.2.3.2 Supported liquid membranes and three-phase extraction

Various approaches have been applied with the aim of intensifying the extraction kinetics. Due to the very low solubility in water of the hydrophobic ILs, supported liquid membranes can be created. Thus, $[P_{6,6,6,14}]$[Phos], which has solubility of 9 mg dm^{-3}, was inserted in supported liquid membranes [60]. A polymer matrix impregnated with pure IL or IL/dodecane solution was used to carry out the removal of butyric acid from its aqueous solution [68]. Kinetics with diluted IL impregnated particles was faster achieving 95% of equilibrium saturation in less than 6 min. Re-extraction was found to be the slower process, due to the tardy decomposition of the acid–IL complexes at the stripping interface. The IL impregnated resin showed good long-term stability in multiple cycles of extraction and stripping.

Pertraction or three-phase extraction is another well-known tool for kinetics intersification. The increase in rate is due to the coupled transport between the two interfaces where two concentration gradients exist [69]. Results about the pertraction of lactic acid through supported liquid membrane showed remarkable (one week) stability of the membrane in continuous operation [70]. Reactive transport of butyric acid through the IL-supported membrane was found to be five times faster compared to the lactic acid [71]. The higher distribution coefficient (multiple acid molecules per one IL molecule) and probably the faster decomposition of the IL–butyric acid complexes are the reasons that could explain this behavior.

The majority of ILs that extract organic acids are highly viscous in dry state. The viscosity, however, essentially drops in magnitude after the IL's saturation with water. Therefore, ILs can be effectively used as bulk membranes in three-phase extractions where recovery and stripping are carried out simultaneously.

7.3 Concluding remarks and challenges of the experimental blanks

The application of liquid–liquid extraction and the related use of biphasic systems have been already validated for recovery, concentration and purification of some fermentatively derived products. Several advantages over the classic separation schemes (ex. chromatographic ones) ease the performance of liquid–liquid extraction in industry. Biphasic systems are designed to be easily handled and to achieve fast separation. They have scale-up potential, require low-cost investments and are suitable for continuous operation mode.

Classical solvent extraction has already become conventional for the antibiotics recovery and may be implemented in organic acids recovery, too, but concerning the bioorganic production of proteins/enzymes, the classical organic solvents satisfy neither the selectivity nor the preservation of the biomolecule functional properties. This has evoked the development of next classes of organic solvents suitable to face the widespread commercialization of fermentatively derived products. Surfactant-based reversed micelle solvents, which allow nano-sized water droplets to be dispersed in a hydrophobic solvent, came to improve the lodging of the guest biomolecule in a safety hydrophilic compartment, offering at the same time new occasions to achieve better selectivity and higher yields [7]. Reversed micelles upgrade the supramolecular organization of the organic solvent thus allowing the accommodation of the biostructures but they do not contribute to the improvement of the solvent from toxicological and ecological standpoint. Here the generation of "green solvents" comes to alter the environmental characteristics of the liquid extractants. ILs represent "designer solvents" that allow great variety of different cations, anions and alkyl chains to be conjugated. In addition to better environmental issue by obtaining non-volatile solvents with negligible vapor pressure, the large number of possible combinations suggests a vast range of polarities that could be acquired. Moreover, the solvent properties could be additionally tuned by coupling ILs with other phase-forming compounds. This constitutes one of the major advantages of the IL-assisted biphasic systems [72]. IL-based systems offer flexible routes for the solute extraction through different types of interactions with the solvent (H-bonding, Coulombic, hydrophobic forces) which are adjusted by the system composition, pH and water content.

From the cited literature, it emerges that the hydrophobic ILs better suit the systems to the requirements of the enzyme and organic acid partitioning. ILs with longer alkyl chains on the cation/anion mildly interact with enzymes and show no disposition to strip off the hydration water from the enzyme surface thus preserving the conformation from unfolding. Moreover, ILs with hydrophobic bulky cations might contribute to an enhancement of the biocatalytic action [22, 23]. Hydrophobic cation tends to decrease the Coulombic interactions and increases the dispersive forces between the hydrophobic proteins and the IL in the IL-rich phase. On this basis segregation and purification of fermentative lipase enzyme have been achieved in the salt-rich phase of the ABS [13, 14]. The proteins that were discarded in the IL-rich phase were of great excess compared with the lipase enzyme, but of the similar molecular weight, which makes them difficult to be removed by conventional ABS. The ILs with quaternary ammonium or phosphonium cations are known to have the capability to self-aggregate in micelles. Micelle formation enables the electrostatic and hydrophobic interactions between the biomolecule and the IL.

ILs with surface-active nature have proved effective for enzyme purification when added as adjuvants to polymeric ABS [17]. In general, the differences in the recovery efficiencies between the PEG-based ABS with and without IL adjuvant are small but they look remarkable on the scale of purification grade. The approach without pre-purification step represents the alternative downstream processing for industrial sectors which do not require a high purity of the outcome enzymes. The adjuvant systems are recommended from a practical viewpoint taking into account the price of the ILs which still remains too high. The adjuvants used are selected among the most common, and thus cheap, imidazolium halides, and are applied in small quantities of 5% of the system weight.

The question of toxicity is compulsory in fermentation production. Some of the ILs employed are not consistent with the requirements for biocompetitiveness and biodegradability [73, 74]. For example, the commonly used $[P_{6,6,6,14}]Cl$ is considered toxic in aquatic environment exhibiting a much higher ecotoxicity than the ordinary organic solvents [75]. The need for novel extractants of improved ecological and toxicological characteristics can still be put forward. Considering the treatment of aqueous streams and bioorganics, the ILs' benign impact on the environment should be convincingly validated in the future experiments. As a whole, the ILs with both hydrophobic anion and cation are expected to exhibit lower ecotoxicity than those comprising both water soluble ions. Ambi-hydrophobic ILs with $[N_{1,n,n,n}]$ (n = 6, 8 or 10) or $[P_{6,6,6,14}]$ cationic moiety and phosphinate or carboxylate anion show great potential as extractants for carboxylic acids' recovery [51–54]. This may give rise to "in situ" extraction when the accumulation of the product (such as organic acid) inhibits the fermentation process.

Re-extraction of fermentative products separated by IL-assisted biphasic systems is another issue that needs to be addressed in the future studies. Enzymes, due to their polymeric structure with high molecular weight, can be subjected to separation from the phase-forming components by a membrane-based (dialysis) technique [17, 21, 25]. The quality of the outcome is considered fairly satisfactory to meet the standards of industrial sectors which do not require a high purity of the enzymes. If further purification is still needed there are not yet experimental studies showing how to proceed. The methods described for recovery and recycling of ILs are not yet integrated in the schemes of bioproducts' separation. Along with the common methods, such as distillation, extraction, adsorption, induced phase separation [76], the membrane-based processes show a promising pathway down to the separation of non-volatile fermentative products from ILs.

Membrane separation is a commercially available and well-studied technology which has successful applications in various industries [77, 78]. Lately this technique has been studied for the purpose of concentrating ILs in order to clean or recover them [76, 79–81].

The nature of the organic acids gives options of choosing methods, other than the membrane-based ones, for the product re-extraction. pH-swing is one of the simplest methods. It was possible, for example, to strip 71% of succinic acid extracted in the top phase of the ABS $[C_6C_1im]Br/(NH_4)_2SO_4$ at pH = 3.43 simply by adding NaOH as precipitant [55]. Successful re-extraction of 95% of L-lactic acid extracted in the IL-rich phase of the ABS $[C_8C_1im][Sac]/H_2O$ at pH = 3.16 was carried out by switching the pH with the addition of an alkaline kosmotrope, K_2HPO_4 [34]. This method has proved effective in stripping acids extracted by ILs *via* reactive mechanism. It was shown that butyric and L-lactic acids are quantitatively recovered from $[N_{1,n,n,n}]$[Phos] or $[P_{6,6,6,14}]$[Phos] by alkali stripping [52, 53, 60]. The re-extraction reported elsewhere [54], however, showed worst results.

Another method suitable for the case of mixed IL/hydrocarbon solvent is molecular distillation. Both the acid and the hydrocarbon can be recovered in their molecular form [66]. It was shown that butyric acid yield in the distillates was 88% and the acid condensed at the temperature of cooling water. A significant portion of 90% of the hydrocarbon (dodecane) was also evaporated. Low-pressure distillation is an option to avoid thermal decomposition of the substances. It was applied to regenerate organic acids extracted in biphasic systems IL/water but the results were discouraging [54].

The outstanding green solvent $scCO_2$ has been recently announced as a new medium for reactive extraction of carboxylic acids [82]. ILs and $scCO_2$ have interesting interactions [83] and their combination could give rise to new solvent properties. On this basis, for example, $[P_{6,6,6,14}]$[Phos] was modified by sparging pressurized CO_2 and a higher extractability of acetic acid was achieved [64].

All said above briefly marks some opportunities ahead and the experimental blanks to be filled in future fundamental work on the liquid–liquid equilibria of the IL-assisted biphasic systems involving fermentative products.

Regarding the process intensification and the equipment design, novel approaches could also be implemented. Minituarization seems a promising one. Microreaction technology is nowadays one of the most innovative and rapid-developing fields in chemical engineering [84, 85]. Microfluidic devices have shown a huge potential in providing high surface-to-volume ratio, higher yield over shorter periods of time, higher product purity and better process control along with reduction in all the expenses associated with the downstream processing.

The major question stuck to the ILs still remains their high price which hinders the industrial use. This may well change in the future considering the ILs' unique potential for recycling and reuse and after improving their ecological and toxicological characteristics when the adoption of “in situ” recovery by ILs will become a feasible solution to the downstream processing.

List of abbreviations

IL’s cationic moiety:	
[Chol]	cholinium
$[C_nC_1im]$	1-alkyl-3-methylimidazolium
$[N_{1,n,n,n}]$	trialkylmethylammonium
$[P_{n,n,n,n}]$	alkyl(tributyl/trihexyl)phosphonium
IL’s anionic moiety:	
Ac	acetate
[BES]	2-[bis(2-hydroxyethyl)amino]ethanesulfonate
$[BF_4]$	tetrafluoroborate
[BIT]	bitartrate
Br/Cl	bromide/chloride
$[C_nSO_4]$	alkylsulfate
[Dec]	decanoate
$[N(CN)_2]$	dicyanamide
[Phos]	bis(2,4,4-trimethylpentyl)phosphinate
[Sac]	saccharinate (which is a benzoic sulfimide)
[TAPSO]	n-[tris(hydroxymethyl)methyl]-3-amino-2-hydroxypropanesulfonate
$[Tf_2N]$	bis(trifluoromethylsulfonyl)imide
Miscellaneous:	
ABS	aqueous biphasic system(s)
CGTase	cyclodextrin glycosyltransferase
IL	ionic liquid
PEG	polyethylene glycol
pI	enzyme isoelectric point
pK_a	acid dissociation constant
PPG	polypropylene glycol
RM	reversed micelle(s)
$scCO_2$	supercritical carbon dioxide
THF	tetrahydrofuran
TOA	trioctylamine

References

[1] Freemantle M. An introduction to ionic liquids. UK, The Royal Society of Chemistry: Cambridge; 2009.

[2] Erythropel HC, Zimmerman JB, de Winter TM, Petitjean L, Melnikov F, Lam CH, et al. The green chemisTREE: 20 years after taking root with the 12 principles. Green Chem. 2018;20:1929–61.

[3] Bogdanov MG. Chapter 7 – ionic liquids as alternative solvents for extraction of natural products. In: Chemat F and Vian M, editors. Alternative solvents for natural products

extraction. Green chemistry and sustainable technology. Berlin, Heidelberg: Springer, 2014:127–66. DOI:10.1007/978-3-662-43628-8_7.

[4] Lozano P, Bernal JM, García-Verdugo E, Vaultier M, Luis SV. Chapter 3 – biocatalysis in ionic liquids. In: Dupont J, Itoh T, Lozano P, Malhotra SV, editors. Environmentally friendly syntheses using ionic liquids, 1st ed. Boca Raton, FL, USA: CRC Press, Taylor & Francis Group, LLC, 2015:31–66.

[5] Tonova K. Separation of poly– and disaccharides by biphasic system based on ionic liquids. Sep Purif Technol. 2012;89:57–65.

[6] Tonova K, Lazarova Z, Nemestóthy N, Gubicza L, Bélafi–Bakó K. Lipase–catalyzed esterification in a reversed micellar reaction system. Chem Ind Chem Eng Quar. 2006;12:175–80.

[7] Tonova K, Lazarova Z. Reversed micelle solvents as tools of enzyme purification and enzyme–catalyzed conversion. Biotechnol Adv. 2008;26:516–32.

[8] Deive FJ, Rodríguez A, Pereiro AB, Jmm A, Longo MA, Coelho MAZ, et al. Ionic liquid–based aqueous biphasic system for lipase extraction. Green Chem. 2011;13:390–96.

[9] Zhao H. Methods for stabilizing and activating enzymes in ionic liquids – a review. J Chem Technol Biotechnol. 2010;85:891–907.

[10] Ulbert O, Bélafi-Bakó K, Tonova K, Gubicza L. Thermal stability enhancement of Candida rugosa lipase using ionic liquids. Biocatal Biotransform. 2005;23:177–83.

[11] Deive FJ, Rodríguez A, Rebelo LPN, Marrucho IM. Extraction of Candida antarctica lipase A from aqueous solutions using imidazolium–based ionic liquids. Sep Purif Technol. 2012;97:205–10.

[12] Hadzir MH, Abbasiliasi S, Ariff AB, Yusoff SB, Ng HS, Tan JS. Partitioning behavior of recombinant lipase in Escherichia coli by ionic liquid–based aqueous two-phase systems. RSC Adv. 2016;6:82571–80.

[13] Ventura SPM, Sousa SG, Freire MG, Serafim LS, Lima ÁS, Coutinho JAP. Design of ionic liquids for lipase purification. J Chromatogr B: Anal Technol Biomed Life Sci. 2011;879:2679–87.

[14] Ventura SPM, Rlf DB, Barbosa JMP, Soares CMF, Lima ÁS, Coutinho JAP. Production and purification of an extracellular lipolytic enzyme using ionic liquid–based aqueous two–phase systems. Green Chem. 2012;14:734–40.

[15] Barbosa JMP, de Souza RL, de Melo CM, Fricks AT, Soares CMF, Lima ÁS. Biochemical characterisation of lipase from a new strain of Bacillus sp. ITP-001. Quím Nova. 2012;35:1173–8. DOI:10.1590/S0100-40422012000600020.

[16] Barbosa JMP, Souza RL, Fricks AT, Zanin GM, Soares CMF, Lima ÁS. Purification of lipase produced by a new source of Bacillus in submerged fermentation using an aqueous two–phase system. J Chromatogr B. 2011;879:3853–58.

[17] Souza RL, Ventura SPM, Soares CMF, Coutinho JAP, Lima ÁS. Lipase purification using ionic liquids as adjuvants in aqueous two-phase systems. Green Chem. 2015;17:3026–34.

[18] De Souza RL, Campos VC, Ventura SPM, Soares CMF, Coutinho JAP, Lima ÁS. Effect of ionic liquids as adjuvants on PEG–based ABS formation and the extraction of two probe dyes. Fluid Phase Equilibr. 2014;375:30–36.

[19] Blesic M, Marques MH, Plechkova NV, Seddon KR, Rebelo LPN, Lopes A. Self-aggregation of ionic liquids: micelle formation in aqueous solution. Green Chem. 2007;9:481–90.

[20] Najdanovic–Visak V, Canongia Lopes JN, Visak ZP, Trindade J, Rebelo LPN. Salting-out in aqueous biphasic systems and salt precipitation. Int J Mol Sci. 2007;8:736–48.

[21] Ventura SPM, Coutinho JAP. Chapter 3 – lipase production and purification from fermentation broth using ionic liquids. In: Xu X, Guo Z, Cheong L-Z, editors. Ionic liquids in lipid processing and analysis: opportunities and Challenges. San Diego, CA, USA: Academic Press and AOCS Press, 2016:59–97. DOI:10.1016/B978-1-63067-047-4.00003-9.

[22] Lee SY, Fa V, E Silva FA, Te S, Taha M, Khoiroh I, et al. Evaluating self-buffering ionic liquids for biotechnological applications. ACS Sustainable Chem Eng. 2015;3:3420–28.

[23] Dreyer S, Kragl U. Ionic liquids for aqueous two–phase extraction and stabilization of enzymes. Biotechnol Bioeng. 2008;99:1416–24.

[24] Souza RL, Lima RA, Coutinho JAP, Soares CMF, Lima ÁS. Aqueous two-phase systems based on cholinium salts and tetrahydrofuran and their use for lipase purification. Sep Purif Technol. 2015;155:118–26.

[25] Lee SY, Khoiroh I, Ling TC, Show PL. Enhanced recovery of lipase derived from Burkholderia cepacia fermentation broth using recyclable ionic liquid/polymer–based aqueous two-phase systems. Sep Purif Technol. 2017;179:152–60.

[26] Ng HS, Ooi CW, Show PL, Tan CP, Ariff A, Mokhtar MN, et al. Recovery of Bacillus cereus cyclodextrin glycosyltransferase using ionic liquid–based aqueous two–phase system. Sep Purif Technol. 2014;138:28–33.

[27] Tonkova A. Bacterial cyclodextrin glucanotransferase. Enz Microb Technol. 1998;22:678–86.

[28] Ng HS, Tan CP, Chen SK, Mokhtar MN, Ariff A, Ling TC. Primary capture of cyclodextrin glycosyltransferase derived from Bacillus cereus by aqueous two–phase system. Sep Purif Technol. 2011;81:318–24.

[29] Ng HS, Tan CP, Mokhtar MN, Ibrahim S, Ariff A, Ooi CW, et al. Recovery of *Bacillus cereus* cyclodextrin glycosyltransferase and recycling of phase components in an aqueous two-phase system using thermo-separating polymer. Sep Purif Technol. 2012;89:9–15.

[30] Aziz NFHA, Abbasiliasi S, Ng HS, Phapugrangkul P, Bakar MHA, Tam YJ, et al. Purification of β-mannanase derived from *Bacillus subtilis* ATCC 11774 using ionic liquid as adjuvant in aqueous two-phase system. J Chromatogr B. 2017;1055–1056:104–12.

[31] Vahidnia M, Pazuki R, Abdolrahimi S. Impact of polyethylene glycol as additive on the formation and extraction behavior of ionic–liquid based aqueous two-phase system. AIChE J. 2016;62:264–74.

[32] Cláudio AFM, Ferreira AM, Shahriari S, Freire MG, Coutinho JAP. Critical assessment of the formation of ionic-liquid-based aqueous two-phase systems in acidic media. J Phys Chem B. 2011;115:11145–53.

[33] Tonova K, Bogdanov MG. Partitioning of α-amylase in aqueous biphasic system based on hydrophobic and polar ionic liquid: enzyme extraction, stripping and purification. Sep Sci Technol. 2017;52:812–23.

[34] Tonova K, Svinyarov I, Bogdanov MG. Hydrophobic 3-alkyl-1-methylimidazolium saccharinates as extractants for L-lactic acid recovery. Sep Purif Technol. 2014;125:239–46.

[35] Kim J-K, Shin O-S, Lee S-J. Optimal conditions for pigmentation in *Bacillus licheniformis* SSA3 and cloning of a DNA fragment involved in pigment production. J Microbiol Biotechnol. 1995;5:22–25.

[36] Lazarova Z, Tonova K. Integrated reversed micellar extraction and stripping of α-amylase. Biotechnol Bioeng. 1999;63:583–92.

[37] Porfiri MC, Picó G, Romanini D, Farruggia B. *Aspergillus oryzae* alpha-amylase partition in potassium phosphate-polyethylene glycol aqueous two-phase systems. Int J Biol Macromol. 2011;49:7–13.

[38] Tonova K, Lazarova Z. Influence of enzyme aqueous source on RME-based purification of α-amylase. Sep Purif Technol. 2005;47:43–51.

[39] Tonova K, Svinyarov I, Bogdanov MG. Ionic liquids–based biphasic systems for enzyme extraction: preliminary data from ionic liquids' screening. J Int Sci Publ: Mat Meth Technol. 2015;9:442–51.

[40] Wu C-Z, Wang -J-J, Li Z-Y, Jing J, Wang H-Y. Relative hydrophobicity between the phases and partition of cytochome-c in glycine ionic liquids aqueous two-phase systems. J Chromatogr A. 2013;1305:1–6.
[41] Pei YC, Li ZY, Liu L, Wang JJ, Wang HY. Selective separation of protein and saccharides by ionic liquids aqueous two-phase systems. Sci China Chem. 2010;53:1554–60.
[42] Zeng Q, Wang YZ, Li N, Huang X, Ding XQ, Lin X, et al. Extraction of proteins with ionic liquid aqueous two-phase system based on guanidine ionic liquid. Talanta. 2013;116:409–16.
[43] Vicente FA, Malpiedi LP, De Silva FA, Jr A P, Coutinho JAP, Ventura SPM. Design of novel aqueous micellar two-phase systems using ionic liquids as co-surfactants for the selective extraction of (bio)molecules. Sep Purif Technol. 2014;135:259–67.
[44] Gutiérrez–Arnillas E, Deive FJ, Sanromán MA, Rodríguez A. Ionic liquids for the concomitant use in extremophiles lysis and extremozymes extraction. Bioresource Technol. 2015;186:303–08.
[45] Passos H, Trindade MP, Vaz TSM, da Costa LP, Freire MG, Coutinho JAP. The impact of self–aggregation on the extraction of biomolecules in ionic–liquid–based aqueous two–phase systems. Sep Purif Technol. 2013;108:174–80.
[46] Van Rantwijk F, Secundo F, Sheldon RA. Structure and activity of *Candida antarctica* lipase B in ionic liquids. Green Chem. 2006;8:282–86.
[47] Tonova K. Long–term preservation of the α-amylase activity in highly concentrated aqueous solutions of imidazolium ionic liquid. Green Process Synth. 2018;7:106–13. 10.1515/gps-2017–0016.
[48] Matsumoto M, Mochiduki K, Fukunishi K, Kondo K. Extraction of organic acids using imidazolium–based ionic liquids and their toxicity to *Lactobacillus rhamnosus*. Sep Purif Technol. 2004;40:97–101.
[49] Tonova K. State–of–the–art recovery of fermentative organic acids by ionic liquids: an overview. Hung J Ind Chem. 2017;45:41–44.
[50] Blahušiak M, Schlosser Š. Physical properties of phosphonium ionic liquid and its mixtures with dodecane and water. J Chen Thermodyn. 2014;72:54–64.
[51] Schlosser Š, Marták J, Blahušiak M. Specific phenomena in carboxylic acids extraction by selected types of hydrophobic ionic liquids. Chem Pap. 2018;72:567–84.
[52] Blahušiak M, Schlosser Š, Marták J. Extraction of butyric acid with a solvent containing ammonium ionic liquid. Sep Purif Technol. 2013;119:102–11.
[53] Marták J, Schlosser Š. Liquid–liquid equilibria of butyric acid for solvents containing a phosphonium ionic liquid. Chem Pap. 2008;62:42–50.
[54] Oliveira FS, Araújo JMM, Ferreira R, Rebelo LPN, Marrucho IM. Extraction of L-lactic, L-malic, and succinic acids using phosphonium–based ionic liquids. Sep Purif Technol. 2012;85:137–46.
[55] Pratiwi AI, Yokouchi T, Matsumoto M, Kondo K. Extraction of succinic acid by aqueous two-phase system using alcohols/salts and ionic liquids/salts. Sep Purif Technol. 2015;155:127–32.
[56] Tonova K, Svinyarov I, Bogdanov MG. Biocompatible ionic liquids in liquid–liquid extraction of lactic acid: a comparative study. Int J Chem Nucl Mater Metall Eng. 2015;9:526–30.
[57] Li QZ, Jiang XL, Zou HB, Cao ZF, Zhang HB, Xian M. Extraction of short–chain organic acids using imidazolium–based ionic liquids from aqueous media. J Chem Pharm Res. 2014;6:374–81.
[58] Reyhanitash E, Zaalberg B, Kersten SRA, Schuur B. Extraction of volatile fatty acids from fermented wastewater. Sep Purif Technol. 2016;161:61–68.
[59] Andersen SJ, Berton JKET, Naert P, Gildemyn S, Rabaey K, Stevens CV. Extraction and esterification of low–titer short–chain volatile fatty acids from anaerobic fermentation with ionic liquids. Chem Sus Chem. 2016;9:2059–63.
[60] Marták J, Schlosser Š. Extraction of lactic acid by phosphonium ionic liquids. Sep Purif Technol. 2007;57:483–94.

[61] Marták J, Schlosser Š. New mechanism and model of butyric acid extraction by phosphonium ionic liquid. J Chem Eng Data. 2016;61:2979–96.

[62] Wang Y-L, Sarman S, Glavatskih S, Antzutkin ON, Rutland MW, Laaksonen A. Atomistic insight into tetraalkylphosphonium-bis(oxalate)borate ionic liquid/water mixtures. I. Local microscopic structure. J Phys Chem B. 2015;119:5251–64.

[63] Sheridan QR, Schneider WF, Maginn EJ. Anion dependent dynamics and water solubility explained by hydrogen bonding interactions in mixtures of water and aprotic heterocyclic anion ionic liquids. J Phys Chem B. 2016;120:12679–86.

[64] Reyhanitash E, Zaalberg B, Ijmker HM, Kersten SRA, Schuur B. CO_2–enhanced extraction of acetic acid from fermented wastewater. Green Chem. 2015;17:4393–00.

[65] Marták J, Schlosser Š. Phosphonium ionic liquids as new, reactive extractants of lactic acid. Chem Pap. 2006;60:395–98.

[66] Blahušiak M, Š S, Cvengroš J, Marták J. New approach to regeneration of an ionic liquid containing solvent by molecular distillation. Chem Pap. 2011;65:603–07.

[67] Blahušiak M, Schlosser Š, Cvengroš J. Simulation of a new regeneration process of solvents with ionic liquid by short–path distillation. Sep Purif Technol. 2012;97:186–94.

[68] Blahušiak M, Schlosser Š, Marták J. Extraction of butyric acid by a solvent impregnated resin containing ionic liquid. React Funct Polym. 2011;71:736–44.

[69] Boyadzhiev L, Lazarova Z. Chapter 7 Liquid membranes (liquid pertraction). In: Noble RD, Stern SA, editors. Membrane separations technology. Principles and applications. Amsterdam: Elsevier Science B.V, 1995:283–52. DOI:10.1016/S0927-5193(06)80009-4.

[70] Marták J, Schlosser Š, Vlčková S. Pertraction of lactic acid through supported liquid membranes containing phosphonium ionic liquid. J Membr Sci. 2008;318:298–310.

[71] Marták J, Schlosser Š, Blahušiak M. Mass–transfer in pertraction of butyric acid by phosphonium ionic liquid and dodecane. Chem Pap. 2011;65:608–19.

[72] Freire MG, Cláudio AFM, Araújo JMM, Coutinho JAP, Marrucho IM, Canongia Lopes JN, et al. Aqueous biphasic systems: a boost brought about by using ionic liquids. Chem Soc Rev. 2012;41:4966–95.

[73] Siedlecka EM, Czerwicka M, Neumann J, Stepnowski P, Fernández JF, Thöming J. Ionic liquids: methods of degradation and recovery. In: Kokorin A, editor. Chapter 28 in ionic liquids: theory, properties, new approaches. 2011. DOI:10.5772/15463.

[74] Egorova KS, Ananikov VP. Toxicity of ionic liquids: eco(cyto)activity as complicated, but unavoidable parameter for task–specific optimization. ChemSusChem. 2014;7:336–60.

[75] Wells AS, Coombe VT. On the freshwater ecotoxicity and biodegradation properties of some common ionic liquids. Org Process Res Dev. 2006;10:794–98.

[76] Mai NL, Ahn K, Koo Y-M. Methods for recovery of ionic liquids – a review. Process Biochem. 2014;49:872–81.

[77] Mohammad AW, Teow YH, Ang WL, Chung YT, Oatley-Radcliffe DL, Hilal N. Nanofiltration membranes review: recent advances and future prospects. Desalination. 2015;356:226–54.

[78] Tylkowski B, Tsibranska I. Polymer application for separation/filtration of biological active compounds. In: Giamberini M, Jastrzab R, Liou JJ, Luque R, Nawab Y, Saha B, et al., editors. Physical sciences reviews. vol. 2, issue 6. Berlin/ Boston: Walter de Gruyter GmbH, 2017:277–92. DOI:10.1515/psr-2017-0022.

[79] Kröckel J, Kragl U. Nanofiltration for the separation of nonvolatile products from solutions containing ionic liquids. Chem Eng Technol. 2003;26:1166–68.

[80] Wang J, Luo J, Zhang X, Wan Y. Concentration of ionic liquids by nanofiltration for recycling: filtration behavior and modeling. Sep Purif Technol. 2016;165:18–26.

[81] Avram AM, Ahmadiannamini P, Qian X, Wickramasinghe SR. Nanofiltration membranes for ionic liquid recovery. Sep Sci Technol. 2017;52:2098–107.

[82] Djas M, Henczka M. Reactive extraction of carboxylic acids using organic solvents and supercritical fluids: a review. Sep Purif Technol. 2018;201:106–19.
[83] Keskin S, Kayrak-Talay D, Akman U, Hortaçsu Ö. A review of ionic liquids towards supercritical fluid applications. J Supercrit Fluids. 2007;43:150–80.
[84] Keoschkerjan R, Richter M, Boskovic D, Schnürer F, Löbbecke S. Novel multifunctional microreaction unit for chemical engineering. Chem Eng J. 2004;101:469–75.
[85] Ehrfeld W, Hessel V, Löwe H. Microreactors: new technology for modern chemistry. Wiley-VCH Verlag GmbH, 2000. DOI:10.1002/3527601953.

Katalin Belafi-Bako, Gabor Toth and Nandor Nemestothy

8 Application of polymer membranes in downstream processes

Abstract: The purpose of downstream processing in a fermentation technology is the isolation, purification and concentration of the final product. Membrane processes are generally used in these steps. In this paper, the application possibilities of polymer membranes in pressure-driven membrane techniques (microfiltration, ultrafiltration, nanofiltration), pervaporation, dialysis and electrodialysis are presented.

Keywords: membranes, polymer, separation

8.1 Introduction

A fermentation technology is usually divided into two main periods: upstream and downstream. Upstream means the preparatory steps for the fermentation (sterilization, inoculation, adjustment of the fermenter and monitoring, controlling systems, etc.), while the purpose of the downstream processing is the isolation of the final product [1–9].

The general features of the fermentation broths pre-determine the techniques of downstream methods. These are as follows:
- aqueous phase (not organic solvent)
- product concentration is usually low
- multiphase system
- multi-component system

These facts show clearly that designing downstream processing is a difficult task, several different separation techniques may be used. In general the downstream processing has 4 main steps as it is presented in Figure 8.1.

In the first step, the removal of solids from the fermentation broth can be accomplished by filtration, sedimentation and centrifugation. If the product is formed in the cell (intracellular compounds), disruption techniques should be applied, i. e. the cell walls have to be destructed to get out the product. The second step is the isolation, which is followed by purification and concentration, and the last step is the final polishing of the product (mainly used in pharmaceutical industry for high

This article has previously been published in the journal Physical Sciences Reviews. Please cite as: Belafi-Bako, K., Toth, G., Nemestothy, N. Application of polymer membranes in downstream processes *Physical Sciences Reviews* [Online] 2020, 7. DOI: 10.1515/psr-2018-0070

https://doi.org/10.1515/9783110574111-008

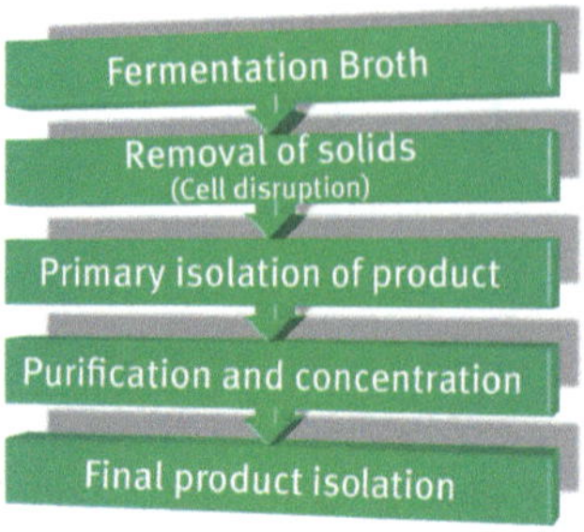

Figure 8.1: General scheme of downstream processing.

purity products). Membrane processes are generally used in the isolation, purification and concentration steps.

The size and complexity of biologically produced substances covers the range from simple organic solutes to macromolecular, colloidal and particulate species [10]. This fact makes the separation tasks extremely difficult and therefore each bioproduct demands its own type of separation. When membrane is going to use in these cases, it is a real disadvantage [11] and makes the process difficult to optimize regarding both process parameters and cleaning.

8.2 Membrane processes in downstream

Membrane separation technologies [12–24] have been applied more and more widely in industrial processes due to their beneficial characters. They can be operated under mild conditions, need and produce no hazardous substances (e. g. organic solvents), the energy consumption is low compared to traditional separation techniques. Thus, membrane separations can be considered as environmental safe, modern processes.

The most well-known membrane separations are the pressure-driven processes: microfiltration, ultrafiltration, nanofiltration and reverse osmosis. Beyond these methods there are other membrane techniques where the driving force is not the pressure difference, but the concentration difference (e. g. dialysis, pervaporation), temperature difference (e. g. membrane distillation) or electric potential difference (e. g. electrodialysis).

8.2.1 Fundamentals of membrane processes

The key element of membrane separations is the membrane itself, which is defined as a permselective barrier [12, 14]. The principle of membrane separation processes is presented in Figure 8.2. During operation the feed stream is divided into two streams: concentrate (the remaining phase) and permeate (passed through the membrane into the other side).

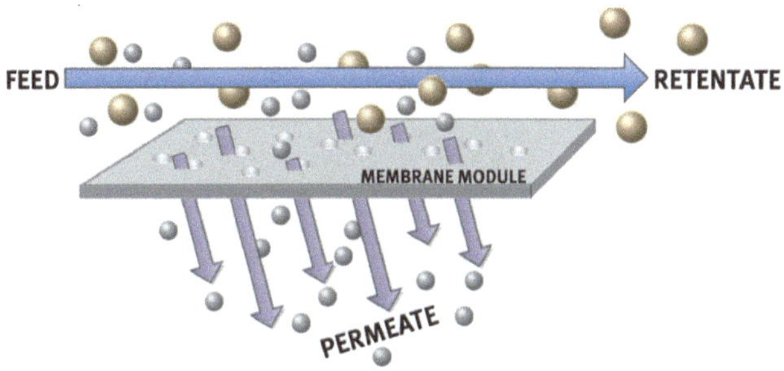

Figure 8.2: The principle of membrane separations.

Membrane separations are characterized by two main parameters: flux and selectivity. Flux is defined as the mass or volume rate permeating through the membrane and it is typically expressed as mass or volume per membrane surface area per unit of time. The selectivity of a membrane can be given by either the retention factor or the separation factor [12].

The membranes used can be classified according to e. g. their phase (solid or liquid), to their origin (natural or synthetic), to their ionic character (neutral or charged), of simply to their material: inorganic, such as metal, ceramic, etc., or organic i. e. polymer. Polymer membranes are applied most widely in industrial technologies, since they are cost-effective, available in various sizes, modules, chemical structures (isotropic, anisotropic, composite, with thin film on top as a selective layer, etc.).

The membranes are usually applied in modular constructions. The basic module types are plate and frame, spiral wound modules (where flat sheet membranes are used), tube and hollow fiber modules. Polymer membranes should be built in a relevant house. The advantage of these types of constructions is the modular character, which makes the scale-up easier.

During the separation process, transport of certain compounds of the feed stream occurs through the membrane. The steps of the transport include passing through various layers (films), which are considered as resistances. These difficulties – including the phenomenon of concentration polarization – should be studied carefully before starting the operation.

Membrane modules can be operated in two modes: dead-end (batch) and crossflow. The batch mode of operation is similar to the traditional tangential filtration, where the direction of the feed is tangential to the membrane and a press cake is formed; thus, filtration should be stopped after a short while, and having cleaned the membrane we start the operation again. On the other hand, in the case of crossflow mode of operation the direction of the feed is parallel to the membrane, and providing sufficiently high flow rate the turbulent flow of the feed stream will sweep continuously the surface of the membrane; thus, press cake formation can be avoided and the system can be operated (semi)continuously.

Although cross-flow operation seems an attractive and efficient method, fouling of the membrane cannot be totally avoided and this is one the most serious problems in membrane technology. In application of membranes in biological matrixes, the situation is even worse due to the complex biological matrix.

8.2.2 Microfiltration, ultrafiltration, nanofiltration

These membrane filtration processes – known as pressure-driven membrane methods – are able to separate according to the size of the solutes. Thus, microfiltration and ultrafiltration are suitable to reject various types of cells, macromolecules, polymers, while nanofiltration can be used for separation of smaller compounds between 200 and 1000 Da.

In some cases microfiltration can be used directly for removal of the cells ("cell harvest") present in the fermentation broth [24, 25]. These mixtures are considered rather *suspensions*, hence their behavior should be studied carefully. For instance yeast cells (*Saccharomyces cerevisie*) are frequently separated by microfiltration after bioethanol fermentation, using various types of membranes. The process has been studied for long and widely, it was even modelled, as well [26].

Beyond yeast cells, other types of microorganisms can be separated by microfiltration or ultrafiltration. Polyamide and polysulfone (100 kDa) ultrafiltration membranes were used successfully to separate the cells (*Lactobacillus bulgaricus*) producing lactic acid from the fermentation broth [27]. After these experiments the fermentation and the separation steps were integrated and a cell-recycle membrane fermentor was established and operated.

Another example is where *Streptomyces rimosus* cells were harvested from the whole fermentation broth (it was an antibiotic producing fermentation – terramycin) by 0.45 μm Durapore membrane cassette in batch operation mode [28]. They found that pH has a significant effect on the performance of the separation. Similarly microfiltration and ultrafiltration was applied to recover spore-crystals from *Bacillus thuringiensis* fermentation broth [29] using 0.22 μm hydrophobic polyvinylidene fluoride and hydrophilic cellulose acetate microfiltration and polyethersulfon ultrafiltration (cut-off 6 and 10 kDa) membranes.

Polish researchers investigated the microfiltration separation of cells from fermented glycerol solution [30], where 1,3-propane-diol was aimed to produce by *Citrobacter freundii* and *L. casei*. Polypropylene capillary modules (Accurel PP S6/2) were used in the experiments with 0.2 and 0.6 μm pore size. They described that strong fouling was experienced during microfiltration therefore a backflushing methodology was developed and applied. Similarly fouling was found as a significant problem in microfiltration of *Polyporus squamosus* fermentation broth [31].

Nanofiltration can be applied for recovery of small molecular weight compounds from the fermentation broth. Typical examples are organic acids which are important targeted compounds in the renewable biorefinery industry [32]. To date a variety of high value organic acids have been produced by microbial fermentation (such as succinic, butyric, fumaric acids), where nanofiltration offers several advantages, e. g. great flexibility, high degree of separation and selectivity, easy integration to other separation units, etc. Various polymer materials are used to prepare nanofiltration membranes including:
- Polypiperazine-amide
- Polyamide
- Poly(vinyl)alcohol/ polyamide
- Aromatic crosslinked polyamide
- Polyethersulfone

Nanofiltration membranes usually have a very thin selective layer and the behavior of this membrane surface (ionic character) toward the feed solution strongly influences the performance of the separation.

As an example, product recovery in lactic acid production by *Bifidobacterium longum* was reported to carry out by nanofiltration [33]. A membrane with a molecular weight cut-off of 100–400 Da was used to separate lactic acid from lactose and cells in the cheese whey fermentation.

8.2.3 Pervaporation

Pervaporation is a special membrane process suitable for separation of volatile compounds. Among fermentation products biofuels: ethanol and other alcohols seem especially good candidates for pervaporative removal.

Biofuel recovery from fermentation systems [34] can be realized by distillation but pervaporation is one of the strongest competitor techniques since it provides advantages such as increased energy efficiency, reduction of capital cost, synergy of performing both alcohol recovery and solvent dehydration, etc.

Another group of volatile compounds is flavor substances, where pervaporation can be applied for recovery. As an example, natural fruity aroma was produced during submerged fermentation by *Pichia fermentans* using sugarcane molasses as a cultivation broth [35]. The aroma compounds were recovered from the fermentation by a pervaporation process using a polydimethylsiloxane membrane (Pervap 4060-Sulzer). Isoamyl acetate, a characteristic compound associated with fruity aromas, was the major compound produced and concentrated from the broth from 9 to 61.8 mg/L at 45 °C. As a single step of downstream operation, pervaporation was efficient for recovering and concentrating the natural fruity aroma, which had a characteristic banana flavor.

In another study the recovery of a wine-must aroma profile, formed by *S. cerevisiae* during a muscatel wine-must fermentation was investigated using organophilic pervaporation [36]. The wine-must samples and the aroma concentrates obtained were analyzed and eight major wine-must components were recognized: four alcohols; three esters; and one monoterpenic compound. In conclusion, it was shown that organophilic pervaporation can be highly suitable for the continuous recovery of very complex and delicate aromatic profiles produced during microbial fermentation.

In some cases pervaporation is suitable for recovery of not too volatile aroma compounds, as well. e. g. gamma-decalactone (4-hidroxydecanoic acid lactone, boiling point is 280 °C) [35] occurring fruit flavors (peach) is produced by fermentation of castor oil in batch concentration of 500 to 1200 ppm, which was recovered and concentrated by pervaporation using polyether-polyamide block copolymer (PEBA) at 40 °C. The flavor compound was possible to enrich significantly. Another similar example is lactone 6-pentyl-2-pyrone, a natural aroma compound of coconut fragrance [37]. It was produced by room temperature cultivation of the fungus *Trichoderma viride*. The concentration of the aroma was below 1000 ppm. Pervaporation directly coupled to the active fermenter yielded product enrichment (and volume reduction) by a factor of 10–20, using PEBA membrane.

8.2.4 Dialysis

Dialysis is a membrane separation technique, where small, unwanted compounds from macromolecules in solution are removed by diffusion [38]. The driving force is the concentration difference between the two sides of the membrane. Porous polymer membranes are normally used, hence the higher macromolecules are retained be the membrane, but small molecules, e. g. buffer salts pass freely through the membrane until an equilibrium is reached. Regarding membrane material, regenerated cellulose, synthetically modified cellulose or synthetic polymers, such as polysulfone, polyacrylonitrile or polyamide are used for dialysis [39]. Although the process is slow and large volume of buffer is required (receiving phase), it is a commonly used method in downstream processes for desalination and buffer exchange of proteins.

8.2.5 Electrodialysis

Electrodialysis is membrane technique, suitable for separation of charged compounds. It uses charged membranes which are either cation or anion selective (Figure 8.3). One of the most attractive groups for electrodialysis separation is acidic compounds, which can be manufactured by fermentation.

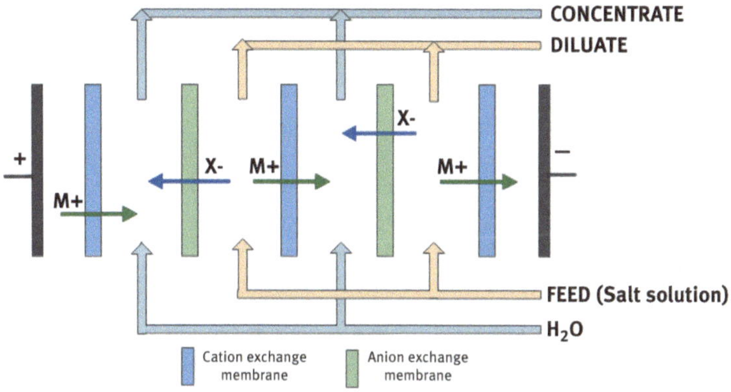

Figure 8.3: Fundamentals of electrodialysis.

As an example, malic acid can be produced by microbial transformation of fumaric acid using *Brevibacterium ammoniagenes, B. flavum or S. cerevisiae* strains. Having removed the cells, the acid can be separated from the fermentation broth by electrodialysis [40].

Other acids can be separated similarly, such as acetic acid, lactic acid, citric acid, itaconic acid, etc. Electrodialysis can be even integrated into the fermentation system; thus, *in situ* product removal can be established.

8.3 Conclusions

Application possibilities of polymer membranes in downstream processing were presented in this paper. Pressure-driven membrane techniques (microfiltration, ultrafiltration, nanofiltration), pervaporation, dialysis and electrodialysis can be used for isolation of various compounds from fermentation broths, also for concentration and purification of certain substances.

References

[1] Flickinger MC. Downstream industrial biotechnology: recovery and purification. Hoboken, New Jersey, USA: John Wiley & Sons; 2013.
[2] Doran PM. Bioprocess engineering principles. 2nd ed. Waltham, Massachusetts, USA: Academic Press, 2013.
[3] Goldberg E. Handbook of downstream processing. London, UK: Blackie Academic & Professional, 1997.
[4] Kumar A, Awasthi A. Bioseparation engineering. New Delhi, India: International Publishing House, 2009.

[5] Ratledge C, Kristiansen B. Basic biotechnology. 3rd ed. New York, USA: Cambridge University Press, 2006.
[6] Wesselingh JA, Krijgsman J. Biotechnology, downstream. Netherlands: Delft Academic Press, 2016.
[7] Okafor N, Okeke CB. Modern industrial microbiology and biotechnology. 2nd ed. Boca Raton, Florida, USA: CRC Press, 2017.
[8] Doble M. Principles of downstream techniques in biological and chemical processes. Boca Raton, Florida, USA: CRC Press, 2016.
[9] Goldberg E. Handbook of downstream processing, blackie academic & professional. London, UK: Chapman & Hall, 1997.
[10] Matson SL. Membrane bioseparations. In: Noble RD, Stern AS, editors. Membrane science and technology. Harvard, USA: Arete Technologies Inc. Elsevier Science, 1995:353–413.
[11] Persson A Downstream processing using membrane technology. PhD thesis, Lund University, 2001.
[12] Mulder M. Basic princinpes of membrane technology. Dordrecht, Netherlands: Kluwer Academic Publisher, 1996.
[13] Cui ZF, Muralidhara HS. Membrane technology a practical guide to membrane technology and applications in food and bioprocessing. Burlington, USA: Elsevier, 2010.
[14] Baker RW. Membrane technology and applications. Boca Raton, Florida, USA: CRC Press, 2012.
[15] Hilal N, Ismail A, Wright C. Membrane fabrication. Boca Raton, Florida, USA: CRC Press, 2015.
[16] Purkait MK, Singh R. Membrane technology in separation science. Boca Raton, Florida, USA: CRC Press, 2017.
[17] Peinemann KV, Pereira Nunes S. Membrane technology volume 1: membranes for life sciences. Strauss Gmbh: Wiley Verlag; 2008.
[18] Peinemann KV, Pereira Nunes S, Giorno L. Membrane technology volume 3: membranes for food applications. Strauss Gmbh: Wiley Verlag, 2008.
[19] Mohanty K, Purkait M. Membrane technologies and applications. Boca Raton, Florida, USA: CRC Press, 2012.
[20] Hoek EM, Tarabara VV. Encyclopedia of membrane science and technology. NY, USA: Wiley, 2013.
[21] Charcosset C. Membrane processes in biotechnology and pharmaceutics. Oxford,UK: Elsevier, 2012.
[22] Field R, Bekassy-Molnar E, Lipnizki F, Vatai G. Engineering aspects of membrane separation and application in food processing. Boca Raton, Florida, USA: CRC Press, 2017.
[23] Crespo JG, Böddeker Karl W. Membrane processes in separation and purification. Dordrecht, Netherland: Kluwer Academic, 1993.
[24] van Reis R, Zydney A. Membrane separations in biotechnology. Curr Op Biotechnol. 2001;12:208–11.
[25] Belfort G, Davis RH, Zydney AL. The behaviour of suspensions and macromolecular solutions in crossflow microfiltration. J Membr Sci. 1994;96:1–58.
[26] Hwang KJ, Ku CY. Model development for estimating microfiltration perforamnce of biotechanol fermentation broth. J Taiwan Inst Chem Eng. 2014;45:1233–40.
[27] Giorno L, Chojnacka K, Donato L, Drioli E. Study of a cell-recycle membrane fermentor for the production of lactic acid by *Lactobacillus bulgaricus*. Ind Eng Chem Res. 2001;41:433–40.
[28] Hooper LA, Hollein HC, Slater CS. Microfiltration of *Streptomyces rimosus*: cell harvesting process studies. Sep Sci Technol. 1998;33:1747–65.

[29] Marzban R, Saberi F, Shirazi MM. Microfiltration and ultrafiltration of Bacillus thuringinesis fermentation broth: membrane performance and soper-crystal recovery approaches. Braz J Chem Eng. 2016;33:88–95.
[30] Gryta M, Tomczak W. Microfiltration of post-fermentation broth with backflushing membrane cleaning. Chem Pap. 2015;69:544–52.
[31] Kristic DM, Markoc SL, Tekic MN. Membrane fouling during cross-flow microfiltration of *Polyporus squamosus* fermentation broth. Biochem Eng J. 2001;9:103–9.
[32] Zaman NK, Law JY, Chai PV, Rohani R, Mohammad AW. Recovery of organic acids from fermentation broth using nanofiltration technologies: a review. J Phys Sci. 2017;28:85–109.
[33] Li Y, Shahbazi A, Williams K, Wan C. Separate and concentrate lactic acid using combination of nanofiltration and reverse osmosis membranes. Appl Biochem Biotechnol. 2008;147:1–9.
[34] Vane LM. A review of pervaporation for product recovery from biomass fermentation process. J Chem Technol Biotechnol. 2006;80:603–29.
[35] Rossi SC, Medeiros AB, Weschenfelder TA, de Paula Scheer A, Soccol CR. Use of pervaporation process for the recovery of aroma compounds produced by P. fermentans in sugarcane molasses. Bioprocess Biosyst Eng. 2017;40:959–67.
[36] Schäfer T, Bengtson G, Pingel H, Böddeker KW, Crespo JP. Recovery of aroma compounds from a wine-must fermentation by organophilic pervaporation. Biotechnol Bioeng. 1999;62:412–21.
[37] Bödekker KW. Recovery of volatile bioproducts by pervaporation. In: Crespo J, Böddeker KW editors, Membrane processes in separation and purification. Springer Science & Business. Dordrecht, Netherland: Kluwer Academic Publishers, 1994.
[38] Baroux E, Nath A, Whitney S. Bioprocess technology II dialysis and reverse osmosis, CHEM-E3140 report. 2016.
[39] Luo J, Wu C, Xu T, Wu Y. Diffusion dialysis - concept, principle and applications. J Memb Sci. 2010;366:1–16.
[40] Bélafi-Bakó K, Nemestóthy N, Gubicza L. Study on application of membrane techniques in bioconversion of fumaric acid to L-malic acid. Desalin. 2004;162:301–6.

Index

https://doi.org/10.1515/9783110574111-009